Data-Based Methods for Materials Design and Discovery

Basic Ideas and General Methods

Synthesis Lectures on Materials and Optics

Data-Based Methods for Materials Design and Discovery: Basic Ideas and General Methods

Ghanshyam Pilania, Prasanna V. Balachandran, James E. Gubernatis, and Turab Lookman
2019

Data-Based Methods for Materials Design and Discovery: Basic Ideas and General Methods
Ghanshyam Pilania, Prasanna V. Balachandran, James E. Gubernatis, and Turab Lookman

ISBN: 978-3-031-01255-6   paperback
ISBN: 978-3-031-02383-5   ebook
ISBN: 978-3-031-00247-2   hardcover

DOI 10.1007/978-3-031-02383-5

A Publication in the Springer series
SYNTHESIS LECTURES ON MATERIALS AND OPTICS

Lecture #1
Series ISSN
ISSN pending.

Data-Based Methods for Materials Design and Discovery

Basic Ideas and General Methods

Ghanshyam Pilania
Los Alamos National Laboratory, Los Alamos, New Mexico

Prasanna V. Balachandran
University of Virginia, Charlottesville, Virgina

James E. Gubernatis
Santa Fe, New Mexico

Turab Lookman
Santa Fe, New Mexico

SYNTHESIS LECTURES ON MATERIALS AND OPTICS #1

ABSTRACT

Machine learning methods are changing the way we design and discover new materials. This book provides an overview of approaches successfully used in addressing materials problems (alloys, ferroelectrics, dielectrics) with a focus on probabilistic methods, such as Gaussian processes, to accurately estimate density functions. The authors, who have extensive experience in this interdisciplinary field, discuss generalizations where more than one competing material property is involved or data with differing degrees of precision/costs or fidelity/expense needs to be considered.

KEYWORDS

materials representations, databases, materials design, materials discovery, multi-objective learning, multi-fidelity learning

Contents

Preface

A little less than a decade ago, a revolution took place in materials research: A surge in using machine learning methods in the design and discovery of new materials occurred. The authors of this volume participated in this surge in different ways. The two senior authors "joined" the revolution in the sense that they became interested in the possibilities this new data-driven approach to materials discovery offered. The two junior authors were part of the revolution in the sense that their theses and post-doctoral work used these methods. While materials research has lagged other sciences, engineering, social sciences, etc. fields in adopting machine learning, in the past decade it seems machine learning has burst forth upon even many more fields. Just about everyone is using data-driven machine learning or its relatives, data mining, and artificial intelligence, for almost everything imaginable.

In this short volume, we recount our experiences in using machine learning methods from the point of view of sharing with the readers some things we know now that we might say we wished we knew when we started doing this type of research. In many respects, this volume builds upon the recent *Physical Review Materials* article [*Phys. Rev.* B 2, 120301 (2018)] written by two of us. In a larger volume, we can expand on several themes mentioned in that article, give more technical details, and add some additional topics.

In the first chapter, we present a historical perspective and provide some background about No Free Lunch Theorems and the Bias-Variance Tradeoff. These latter two themes are intrinsic to machine learning but are rarely discussed in the materials literature. What they imply is machine learning is largely a heuristic endeavor, where so to speak, "You pay your money and take your chances."

In the hard sciences, including the materials sciences, for over a century, we have fitted mainly experimental data to models. The models are usually analytic forms derived from approximate theory or phenomenological forms based on physical intuition. In machine learning, we are still fitting data to models but the models have no physical basis but rather are proposals to facilitate learning specific types of information. What is part of machine learning is not simply determining the parameters of the models but also selecting which parameters to use. In general, there are two types of parameters. One type are not necessarily independent random variables, called features, attributes, or descriptors, that enable us to distinguish one material from another while at the same time being effective in capturing trends and patterns in the data. The choice of these parameters is where the materials science enters. The others are called hyper-parameters that specify the assumed functional form of the model. These are defined by the machine learning methods we choose to use. What had been common was choosing the features to be variables describing common physical properties, such as ionic radius, electronegativity, valence, etc. of the constituents atoms and ions. As we discuss in Chapter 2 and illustrate in the other chapters, this "elemental" choice is not always the most effective one, let alone to most useful one. In fact, there are so many choices that one problem in the proper application of machine learning methods is reducing the number of features used.

Machine learning requires data. In Chapters 3 and 4, we discuss two scenarios—situations where we have large databases and where we grow larger ones from small data bases. An active research area today is generating large databases, usually by using high throughput theoretical calculations and experimental measurements. We can probe and query these databases about a variety of materials properties and chemical trends and infer potential new materials relative to what might be possible but what is yet observed. For small databases, we seek the same discoveries but in a controlled low throughput mode where based on what we know we add new entries to the database iteratively by procedures that in a well-defined sense improve our current state of knowledge.

It is more precise to say that in materials design and discovery we seek new and enhanced properties of materials instead of simply new materials. It is even more precise to say that we want a material with more than one property enhanced. An issue is improving one property often degrades the other property of interest. How do we balance these opposing trends? Under the rubric of multi-objective optimization, we address such issues in Chapter 5. There, we also discuss several material design and discovery success stories.

In Chapter 6, we discuss multi-fidelity optimization. This is a class of machine learning methods that addresses a practical problem. High precision calculations and measurements, even if possible, are typically time consuming and costly. Multi-fidelity optimization methods allow costly and time-consuming procedures to be combined with less costly and time consuming ones to produce machine learning models whose predictive precision approaches those produced by using only the more expensive and time consuming methods. This seems like something for nothing. In essence, the expectation to get something for nothing is a reason why the community turned to machine learning methods. It is important to bear in mind that machine learning builds statistical models, and statistics in turn is based on probability theory. Probability theory is what we turn to when we cannot determine the answer.

We close by discussing some other potential applications of machine learning in Chapter 7. In keeping with the "multi" themes of Chapters 4, 5, and 6, we note the intriguing existence of multi-relational methods, which as far as we know, have yet been applied to materials science problems.

Having finished summarizing what is in this volume, we conclude by saying a few words about what is not. We only address applications of machine learning methods to cases where our data is static; that is, we do not discuss analyzing data on-the-fly that is changing in real-time dynamic processes or using such methods to control and optimize such processes themselves. We discuss only a selection of common machine learning methods applied to static tables of data. For example, we do not discuss the beautiful and widely popular support vector machine classification and regression methods or the intriguing neural network methods, including the touted deep learning methods. Discussions of these methods are in many textbooks or can otherwise be found by simply searching the Web. We do not discuss unsupervised learning methods, methods that do not use data unaccompanied by a label that specifies a materials class or the value of some functionality. In retrospect, we observe that much of what we discuss belongs to a class of methods based on what are called Gaussian processes. This rooting underscores the probabilistic foundation of machine learning, and hence the basic task of accurately estimating means from complex multi-dimensional density functions and more importantly, for error estimation, the necessity of estimating accurately the dispersion of information about the means. In the appendices we provide some background to probability theory and Gaussian processes.

We hope readers will find this short volume useful for its perspectives and examples of the potential for machine learning as an important approach to the design and discovery of new materials. Its potential for these goals is evolving rapidly.

Ghanshyam Pilania, Prasanna V. Balachandran, James E. Gubernatis, and Turab Lookman
December 2019

Acknowledgments

Our machine learning understanding and work has greatly benefited from conversations and collaborations with J. Hogden, R. Ramprasad, M. Scheffler, J. Teiler, D. Wolpert, and D. Xue. We would also like to thank Jeanine Burke at Morgan & Claypool for her support and encouragement. G.P. further acknowledges support from the Alexander von Humboldt Foundation and Los Alamos National Laboratory's Laboratory Directed Reseach and Development (LDRD) program.

Ghanshyam Pilania, Prasanna V. Balachandran, James E. Gubernatis, and Turab Lookman
December 2019

CHAPTER 1

Introduction

After outlining the scope and plan of the volume, we give a basic introduction to what is driving the burst of interest in using machine learning methods to aid in the design and discovery of new materials, an overview of what is machine learning, and a discussion of what are some of the basic issues in using these methods. With respect to the latter, we explore the bias-variance tradeoff and the No Free Lunch Theorems. Both topics are often not discussed in introductions of machine learning for materials science but are in fact general issues facing all machine learning applications whether for materials research or other applications.

1.1 SCOPE AND PLAN

Using machine learning methods to aid the design and discovery of new materials is a rapid growth area in materials research and will continue to be so for the foreseeable future. At least in the United States, this explosive growth was fueled by President Obama's 2011 announcement of the *Materials Genome Initiative* (MGI) [1] and the ensuing governmental infrastructure developed to coordinate this initiative's implementation by various funding agencies.

The intent of the MGI is to halve the time it takes to discover new materials. It was to impact materials research in manner similar to the way the *Human Genome* Initiative impacted the biological and medical sciences. The focus of this latter initiative was accruing large databases and using advanced statistical methods to learn from the data what is important for understanding how to treat diseases, design new pharmaceuticals, etc. For the materials sciences, the intent is accruing large databases and using advanced statistical methods to learn from the data how to generate new materials with enhanced or novel properties. The hope is to design and discover game-changing materials, those with one or more properties that lie far beyond those currently known. In turn, these new materials would enable the development of new technologies and lead to the marketing of new products based on their extraordinary properties. In other cases, alternative materials are sought, perhaps for reasons of cost or environmental friendliness. In MGI announcements, Kevlar and Li-based batteries are oft-quoted examples of new materials marketed because they have exceptional properties or are desirable replacements.

Certainly, the MGI intent is not a novel idea. For the most part, its goal has always been a goal of materials research. What is somewhat new is the explicit emphasis on assembling large databases and use of computational methods to learn from the data. There is hope and reason to believe that these machine learning methods, when applied to properly crafted materials databases and properly used, will augment and in many cases supplant the time-consuming, intuition-based, trial-and-error experimentation that has been the traditional route to the design and discovery of new materials. We say "somewhat new" to acknowledge that much work preceded the MGI announcement and this activity was already growing internationally. This widespread growth was being fueled by the greater accessibility to large-scale experimental light and neutron sources, bigger, faster, and cheaper computers, cheap data storage, and a variety of powerful and well-maintained open source software. In short, consistent with the MGI and related world-wide activity, the way many materials scientists do research today is much more data- and computation-driven than in the past. Not only are computational methods, such as machine learning methods, being used more widely to learn from the data, but also the computations and large-scale experiments are, in fact, creating large databases.

The intent of this short volume is not to review fully the progress and status of machine learning in materials research but rather to convey "some lessons learned" that illustrate the potential of machine learning methods in several, perhaps not so obvious but in fact essential, ways. We will be conveying things that either we wish we knew from the start or else follow paths that up to now have been less traveled. For example, to date, much of the use of machine learning methods in materials science is strongly coupled to the nearly simultaneous generation of large databases by high-throughput *density functional theory* (DFT) calculations (and in some cases the generation of databases by high throughput experiments). However, examples now exist where the use of less common machine learning tools makes it possible to grow small experimentally generated databases into larger ones and along the way to predict new materials. In Chapter 4, we review this approach and discuss its recent use with non-high throughput DFT calculations: instead of an agenda of computing all possibilities with computational accuracy sacrificed for speed, the alternative approach focuses on identifying a few possibilities at a time that will make the most difference with respect to what we presently know, computing these possibilities as accurately as possible, and verifying them experimentally.

In Chapters 5 and 6, we discuss what are called *multi-objective* and *multi-fidelity* optimization methods, several classes of methods relatively new to machine learning. Multi-objective machine learning methods optimize the search for materials needing the simultaneous enhancement of multiple properties. Whether the databases are large or small, generated experimentally or computationally, the motivation for discovering new materials is generally coupled in an essential way with the need for ones with several specific functionalities more advanced than those we currently have. For example, we might want new shape memory alloys with lower hystereses and higher Curie temperatures. Multi-fidelity optimization methods permit the combination of calculations with different levels of accuracy, experiments with different measurement precisions, and even calculations and measurements together with the resulting precision approaching that of the more expensive calculations or measurements. Examples here include combining many high throughput DFT calculations that use less accurate approximations for the electronic interactions with fewer DFT calculations that use more accurate ones to produce band gap predictions whose accuracies approach an all-high fidelity analysis.

In short, instead of giving a somewhat standard summary of textbook machine learning ideas and methods, we focus on the broader picture, discuss some newer methods, and reference their successes. In the next section of this chapter, we present a brief historical perspective that gives a simple example of how data has been used to search for new materials to show how machine learning allows us to build upon it. We then give a somewhat selective description of machine learning. What is it? It is not physics, chemistry, or materials science. How does this domain knowledge enter? It is largely upon us. As we discuss, there are different types of learning and with each type there are different tasks. Machine learning is about using data as input and selecting a machine learning algorithm to build a model that completes the task. The data and hence the models are parameterized by what are called *features*. Learning which features to use is part of the process.

In ensuing chapters, we discuss several applications of different machine learning methods to the prediction of new materials, note cases where the predictions have been validated experimentally, and thereby illustrate the broad spectrum of possible applications [2]. We only give descriptions of the machine learning algorithms used in our examples. There are a host of standard algorithms being used in other contexts. The field of machine learning itself is supported by a number of textbooks describing these algorithms [3–9]. In addition, there is a growing number of review articles [10–20] describing the use of these standards in materials research.

1.2   HISTORICAL PERSPECTIVE

One of the earliest instances of using data and in particular data generated from theoretical calculations to assist in the design and discovery of new materials was the search for new semiconductors, the game changing materials of the 1970s. A pioneering paper by Chelikowsky and Philips [21] states the vision:

> "Structural energies are, for the most part, too small to be calculated quantum mechanically. ... However, if we consider the problem from the point of view of information theory, then the available structural data already contain a great deal of information. ... Thus, one can reverse the problem, and attempt to extract from the available data quantitative rules for chemical bonding in solids."

In other words, the Periodic Table establishes trends in the chemical properties of the atoms as one moves across its rows. In the solid state, remnants of these trends persist. The problem is to extract these trends from the data and use them to predict new materials. This vision is also that of much of today's use of machine learning in the materials sciences.

These early searches resulted in the introduction into the materials design and discovery process of what are called *structure maps* (for example, Figure 1.1). Initially applied to octet *AB* materials [22], these maps are simply scatter plots of two physical properties of the constituent *A* and *B* atoms, such as their ionization potentials, valences, ionic radii, etc. A pencil and ruler is used to draw boundaries between the plotted data that group known materials with the same

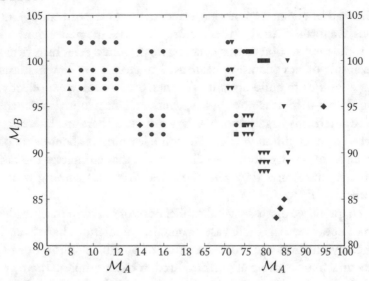

Figure 1.1: Structure map of the octet *AB* compounds with Pettifor's Mendeleev numbers as the co-ordinates. The different symbols shapes denote the different crystal classes: the circles are rock salt; the down-pointing triangles are zinc blende; the squares are wurtzite; the up-pointing triangles are cesium chloride; and the diamonds are diamond. Symbols colored blue mark compounds expected to be ionically bonded (those in rock salt, wurtzite, and cesium chloride structures); red, covalently bonded (zinc blende and diamond). For clarity, bounding boxes clustering the different crystal structures are not drawn.

crystal structure. The challenge was to identify the physical quantities, which in machine learning language are called *features, attributes, or descriptors*, to place on the x and y axes quantities that promoted the greatest segregation of materials into the same structure. "Holes" in the resulting data clusters represented possible new materials and their likely crystal structures. In machine learning language, this segregation problem is called a *classification problem*.

Past work by a number of authors identified a very powerful pair of materials features that classify the octet solids very well. These features are the r_σ and r_π pair first used in a structural plot analysis by St. John and Bloch [23] who chose as the x and y co-ordinates the symmetric combinations

$$r_\sigma = |(r_p^A + r_s^A) + (r_p^B + r_s^B)| \tag{1.1}$$

$$r_\sigma = |r_p^A - r_s^A| + |r_p^B - r_s^B| \tag{1.2}$$

of the s and p orbital dependent radii of the A and B atoms estimated from an early pseudo-potential.

The structure map approach was extended to all *AB* materials with the symmetric combinations of ionic radii but with r_A and r_B computed by a different pseudo-potential [24], and the 574 then known *AB* materials were classified into 34 crystal structures. In the mean time, Pettifor [25] proposed a different set of co-ordinates for a structure map based on what he called *Mendeleev numbers*. This one-dimensional sequence of numbers relabels the elements in the two-dimensional Periodic Table for the most part by going down the columns. He showed doing this captures co-ordination tendencies of the elements and hence structural similarities between materials differing by the presence of one or more of these elements. These co-ordinates do not require computation or measurement and hence became readily used in structure maps for clustering materials with the same physical property other than crystal structure, such as melting temperatures.

In Figure 1.1, we show how the octet solids classify in a structure plot based on Pettifor's Mendeleev numbers. The octets exhibit five crystal structures, with rock salt being the most common. What is difficult about segregating these solids is drawing the boundary between a small set of rock salts, zinc blendes, and wurtzites (particularly between some zinc blendes and wurtzites) whose ground state energy differences are small and positions in a structure plot are close. The few cesium chloride solids sit near but apart from the rock salts. In this figure, we also show what happens if we reduce this multi-class classification problem to a binary one. First, we group the cesium chlorides with the rock salts, then group the remaining three structures. These two groups are principally a separation of these solids into those whose bonding is strongly ionic and those whose bonding ranges from predominantly covalent to very strongly covalent. Although this grouping reduces the number of boundaries machine learning needs to draw, it retains the core difficulty of a classification problem.

What would happen if we were to use simultaneously both the ionic radii and the Mendeleev numbers to capture the two different types of trends? Obviously, we would need more than a ruler and pencil to identify and draw the boundaries between the data clusters because we do not know how to plot anything in four dimensions. In this particular case, machine learning is the replacement for the pencil and ruler: It allows us to extend the concept of a structure map to three or more dimensions, provides us with a method to remove the subjectivity in the decisions of where to draw the boundaries, and replaces boxy boundaries with more sophisticated and flexible multi-dimensional manifolds [26].

While a structure map is useful for a first cut in identifying possible new materials from data, it is clearly limited. It is interesting to note another type of two-dimensional scatter plot, called the *Ashby plot* (Figure 1.2), that is used in the materials engineering problem of selecting the best material to use in a particular application [27]. It displays two properties of many materials for multiple classes of materials simultaneously. Historically, an Ashby plot displayed the logarithm of the Young's modulus E vs. the logarithm of the density ρ for overlays of metals, polymers, ceramics, foams, etc. Specific procedures evolved for choosing the material class and the material in that class for the application at hand, for example, depending on whether the

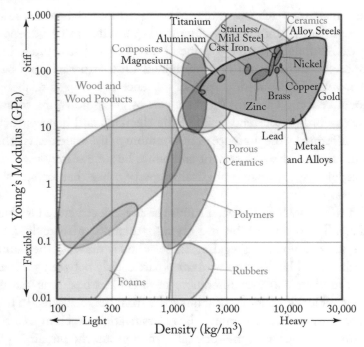

Figure 1.2: An Ashby log-log plot with Young's modulus and density as the co-ordinates. Multiple material classes are represented.

material would be used as a rod in tension or as a plate subject to bending. A typical use of an Ashby plot is material selection in product design. If one has a rod in tension, for example, the Young's modulus E divided by the density ρ is the performance index. If we want $E/\rho \geq k$, then the line $\log E = \log k - \log \rho$ is drawn on the figure. Those materials that lie above the line are the best materials and those below, the worst. While identifying which materials we might choose, it does not identify which specific material we should chose.

What is characteristic for each material class in an Ashby plot are limiting boundaries of high Young's modulus and low density, high Young's modulus and high density, etc. From the point of view of an Ashby plot, the search for new materials is about pushing one of these boundaries in a favorable direction. For example, for alloys of the form $A_x B_y C_{1-x-y}$ with A, B, and C fixed, for what values of x and y will we find a material with higher Young's modulus and lower density relative to the alloys we already know? From a machine learning perspective these boundaries are called *Pareto fronts*. Predicting such fronts in an active subfield of machine learning. As we will discuss later (Chapter 5), machine learning methods exist that adaptively predict values of x and y whose Young's modulus and density are likely to push the front favorably. Used this way, machine learning can proactively address the basic goal in materials design and discovery for product development.

1.3   MACHINE LEARNING

1.3.1   AN OVERVIEW

Most of the natural sciences, including the materials sciences, are somewhat "Johnnies come lately" in their use of machine learning methods. The widespread use of these methods in the engineering sciences, social sciences, financial sciences, statistics, marketing, etc., have led to a plethora of methods and techniques that are application independent, even though each originated in a particular field. For example, we will discuss the use of machine learning in identifying possible new perovskite materials. The gradient tree boosting method used there has been used to study the short-fin eel population in the waters off New Zealand. From this perspective of newness, "What is *machine learning*?" is a natural question to ask.

In his book *Machine Learning: the Art and Science of Algorithms That Make Sense of Data*, Flach [8] says "Machine learning is the systematic study of algorithms and systems that improve their knowledge or performance with experience." Several times he asserts that "Machine learning is concerned with using the right features to build the right models that achieve the right tasks." He further states that "Tasks are addressed by models, whereas learning problems are solved by learning algorithms that produce models." What does all this mean? It is not simple to say. On the one hand, these statements say machine learning is about learning the algorithms that define the space of models for the data. On the other hand, they say it is about using these algorithms to learn the parameters of the models and then make predictions and decisions. In practice, however, machine learning, at least in the materials sciences, seems to be more about being a toolbox of methods for processing data. Data is fed into one of many possible tools and then predictions and decisions are made. The task is to choose an algorithm for the problem at hand that has good theoretical and empirical performance. We assert that machine learning is all of the above.

Machine learning is, in fact, a subfield of statistics and *artificial intelligence*. Statistics studies data by using different analysis techniques and collection methods. Artificial intelligence studies any agent that perceives its environment and takes actions that maximize its chance of successfully achieving its goals. Machine learning constructs or uses various methods to build analytic models of data to infer from the data certain types of information. The models improve as more data is gathered, that is, as more experience is gained. What types of information and what types of methods are applied depends on the type of query is being made. The user and not the learning method chooses the method: The learning method is not an intelligent agent. Machine learning differs from *data mining*, another subfield of statistics and artificial intelligence. In general, machine learning yields predictive models whereas data mining yields descriptive models. The description is often the identification of patterns in the data that are difficult to discern because of the large number of data points or the large number of features associated with each data point. Both artificial intelligence and data mining use methods of machine learning.

In machine learning, there are also a number of different learning types. In the discussion below, we only consider the two most common ones: *supervised learning* and *unsupervised learn-*

ing. Supervised learning algorithms build models from a set of data that contains both the values of the features, such as the electro-negativities of the constituents, lattice constants, etc., and the value of a label for a physical property, such as crystal structure, band gap, Curie temperature, etc. The labels are known verifiable results. The data are sometime called *instances* or *samples*. Unsupervised learning, on the other hand, builds the mathematical model from data that lacks labels. Subcategories of supervised learning include *semi-supervised learning* and *active learning*. In semi-supervised learning, not all the data are accompanied by labels. We will not consider this type of learning further. Active learning generally starts with a limited set of labeled data and optimizes the values of new features for which the user acquires the labels and augments the data before relearning the model. This type of learning is an important part of this book. We discuss and apply it in Chapter 4.

In general, supervised learning produces predictive models and unsupervised learning descriptive ones. The typical tasks addressed by supervised learning include classification, regression, and density function estimation. In *classification problems*, the label is a variable that maps to a small number of integers. In our example of the classification of the *AB* alloys, the labels were the five crystal structures which map to the integers 1–5. In *regression problems*, the label is a continuous real number, for example, the melting temperature of the solid. The most common task for unsupervised learning is *clustering*, identifying subgroups of data for which the members of each subgroup by some measure of similarity are all closer to each other than to members of another subgroup.

In both types of learning, the inputs to the learning algorithm are the data and the algorithm's output is an analytic model. Predictions and descriptions are made from the model. There are at least three kinds of models: geometric, logical and probabilistic. In *geometric models*, some notions of distance and hypersurfaces are present. For example, finding a complex surface was the object of the classification problem in our Historical Perspectives section. Strings of queries such as "if X then Y" define *logical models*. Decisions trees, which we discuss in Chapter 3, are examples of such models. *Probability models* are based on reasoning that is consistent with the basic axions of probability theory. Gaussian process modeling is a prime example. It is the focus of methods discussed in Chapters 4, 5, and 6.

In both types of learning, some form of *cross–validation* [8, 9] quantifies the uncertainty. In cross-validation, part of the data is used to train the model and the other part is to used to test it. In supervised learning, the model is built with labeled training data. Once constructed, it is then tested by comparing its predicted labels against the known labels of the testing data when its features are used as inputs to the construct. The process is trickier for unsupervised learning because the data does not contain labels. Nevertheless, as the inputs are varied, we can study which parts are relatively invariant to change. Most machine learning methods minimize something. Another difficulty with unsupervised learning is that many local minima may accompany the best fit of the models to the data. With unsupervised learning the first problem is ensuring the algorithm has found the global minimum. In this book, we only consider supervised learning algorithms.

There is a host of different schemes to divide the data into training and testing sets; hence, there are many ways to perform cross-validation. A simple one is to divide the data into N groups selected randomly, then in turn hold out one at a time each of the N as the testing set and build the model on the remaining $N - 1$ sets used as the training set. Then, after sequencing through each of the hold-outs, we can compute the average of the predictions. If N is large enough, we can meaningfully compute an estimate of the standard deviation associated with this average.

There are several concerns about applying cross-validation properly. One is simply whether there is enough data so that the data in either the training or testing set is properly representative of the sample space of the materials and properties being studied, that is, do the training and testing sets properly sample the entire distribution of the data or just a portion of it. If we return once again to our example for the octet solids, we first note that there is only a small number of data and for some crystal structures there are just a few data. Breaking this data up could result in the training or testing set lacking a member of one or more structures. Here, it is advantageous to create training and test sets with each having the structures in the same proportion as the original data. Doing this is clearly a more complicated cross-validation procedure but one that is more representative of the sample space of the problem. Overall, cross-validation is a heuristic: It has no formal justification. It can fail.

While machine learning is about learning algorithms that produce models, from the toolbox point of view, the task is not about learning a new algorithm but about choosing an effective one and then validating it. The path to model validation is clearer for supervised learning than it is for unsupervised learning. For supervised learning, cross-validation is *de riguer*. The ultimate validation is based on what does the model predict when presented with an input not in the original data. In the materials sciences, we have the luxury of having the final arbiter generally being new experiments or calculations. Many other fields that use these methods do not have this option. Not surprisingly, they tend to be much more data intensive.

In machine learning, there are many generic tasks [7–9]. These tasks include modeling the data by some probability distribution function, clustering the data, classifying the data, regression analysis, feature reduction, etc. For each task, numerous methods exist. What defines the discipline using any of these methods is not just the source of the data and the information sought from it but also the parametrization of the data, that is, the features in the data. The *features* define the values of the parameters we input into the machine learned models. The features are sometimes called *descriptors* or *attributes*. They are ultimately the main portals for the material science. In Chapter 2 we discuss a number of ways materials are being represented.

Adding a few more words, we note that in machine learning it is implicit that a data point is a sample drawn from some probability distribution function that represents the complete description of the problem. We do not know the independent features that parameterize this distribution. For a given material, on the other hand, we can easily conjure up a host of physical properties relative to the constituents of the material and to the material as a whole that we believe are relevant and use all as features. In the Historical Perspectives section, for example, we noted that different parameterizations of structure maps, that is, different choices of features, changed how well the data separated into crystal structures. In machine learning language, the search for the best structure map was a search for the best feature set. In short, using what we know *a priori* about what physical quantities control the properties of interest is currently where our domain knowledge enters machine learning. Often, we can propose too many features with most not being independent of each other. *Occam's razor* controls our psyche: Less is more. While there are machine learning methods to aid in identifying from this set the ones most and least important, the burden is upon us to choose how we populate the set. We can depopulate or repopulate as needed. Choosing the right and proper number of features is an important part of machine learning. This leads us to addressing what in machine learning is called the bias vs. variance problem.

1.3.2   BIAS vs. VARIANCE PROBLEM

Likely while in high school or thereabouts, we all were told that if you use enough parameters you can fit anything. Indeed, this is the case. In machine learning, this dictum, however, gets trumped by the bias (the error of the fit) vs. variance (the variations in its predictions) problem [8]. In standard applications, machine learning returns a statistics-based model built upon the data. In making predictions, we use this model mainly to interpolate between the data we know. If we fit the data too accurately (low bias), the risk is we become extremely limited in how far we can controllably interpolate (high variance) beyond any fitted data point. In general, experience shows that low bias models have high variance and vice versa. This is the *bias vs. variance problem* [8, 9]. Restricting our discussion, for convenience, to cross-validation of a regression problem, we will now explore this problem more formally. It is central to supervised learning.

Suppose we have a training set of n data $\mathcal{D} = \{(x_i, y_i), | i = 1, \ldots, n\}$ where an x_i is a vector of the values of the D features and the y_i are the labels for material i. We assume this data is sampled from the joint probability density $p(x, y)$. With our data $\mathcal{D}$, we choose some machine learning regressor algorithm and build a regressor model, $Y_{\mathcal{D}}(x)$ that depends on the data and is a function of the values of the features. For a given x, the regressor predicts y. The algorithm by some means fixes the other parameters of the model, called *hyper-parameters*, to the data.

In practice, we do not know $p(x, y)$ and never will. This is part of the problem. To proceed, we assume we know it and also that we can generate an infinite number of training data sets of the same size. Under these assumptions we can calculate the following two quantizes. First, for any given feature vector x there is some distribution for y. Accordingly, we can compute the expected value of a label

$$\bar{y}(x) = E_{y|x}(y) = \int y p(y|x) dy.$$

Since our data is a random variable, so is our model. Accordingly, we can also compute an expected model

$$\bar{Y}(x) = E_{\mathcal{D}}[Y_{\mathcal{D}}(x)] = \int \int Y_{\mathcal{D}}(x) p(\mathcal{D}) d\mathcal{D},$$

where $p(\mathcal{D})$ is the probability density associated with sampling $\mathcal{D}$ from $p(x, y)$.

Given the model, we can now compute the expected testing error of some label y associated with some feature vector x in our testing set. Using the simple mean squared error for simplicity, we have

$$E_{x,y,\mathcal{D}}\left[[Y_{\mathcal{D}}(x) - y]^2\right] = \int \int \int [Y_{\mathcal{D}}(x) - y]^2 \, p(x, y) p(\mathcal{D}) dx dy d\mathcal{D}.$$

If we knew the distribution of the features and labels, this expectation is the quantity that measures the quality of the machine learning . We now rearrange the right-hand equation to unfold several interpretable contributions to it. First, we write

$$
\begin{aligned}
E_{x,y,\mathcal{D}}\left[[Y_{\mathcal{D}}(x) - y]^2\right] &= E_{x,y,\mathcal{D}}\left[\left[[Y_{\mathcal{D}}(x) - \bar{Y}(x)] + [\bar{Y}(x) - y]\right]^2\right] \\
&= E_{x,\mathcal{D}}\left[[Y_{\mathcal{D}}(x) - \bar{Y}(x)]^2\right] + E_{x,y}\left[[\bar{Y}(x) - y]^2\right] \\
&\quad + 2E_{x,y,\mathcal{D}}\left[[Y_{\mathcal{D}}(x) - \bar{Y}(x)][\bar{Y}(x) - y]\right].
\end{aligned}
$$

With a few steps of algebra, we can show that the far right term is zero. Our expression for the mean squared testing error reduces to

$$E_{x,y,\mathcal{D}}\left[[Y_{\mathcal{D}}(x) - y]^2)\right] = E_{x,\mathcal{D}}\left[[Y_{\mathcal{D}}(x) - \bar{Y}(x)]^2\right] + E_{x,y}\left[[\bar{Y}(x) - y]^2\right].$$

The first term on the right-hand side is the variance associated with the predictions of the fitted regressor relative to those of the expected regressor. We can gain some meaning from the far right term by rewriting it as

$$
\begin{aligned}
E_{x,y}\left[[\bar{Y}(x) - y]^2\right] &= E_{x,y}\left[\left[[\bar{Y}(x) - \bar{y}(x)] + [\bar{y}(x) - y)]\right]^2\right] \\
&= E_{x,y}\left[[\bar{y}(x) - y]^2\right] + E_{x}\left[[\bar{Y}(x) - \bar{y}(x)]^2\right] \\
&\quad + 2E_{x,y}\left[[\bar{Y}(x) - \bar{y}(x)][\bar{y}(x) - y]\right].
\end{aligned}
$$

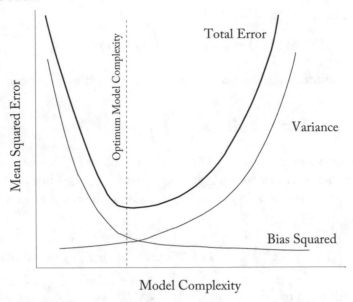

Figure 1.3: Schematic illustration of the testing error as a function of the complexity of the model. Here, noise is absent so the mean squared-error is a combination of the variance and bias squared. Varying the complexity generally produce a model for which the combination of the bias squared and variance combine to give the minimum error. The minimum error generally is not the spot of lowest bias. After [28].

Again, with a few steps of algebra, we can show the far right term is zero. Thus, we have that

$$\underbrace{E_{\boldsymbol{x},y,\mathcal{D}}\left[[Y_{\mathcal{D}}(\boldsymbol{x}) - y]^2\right]}_{\text{Expected Test Error}} = \underbrace{E_{\boldsymbol{x},\mathcal{D}}\left[[Y_{\mathcal{D}}(\boldsymbol{x}) - \bar{Y}(\boldsymbol{x})]^2\right]}_{\text{Variance}}$$

$$+ \underbrace{E_{\boldsymbol{x}}\left[[\bar{Y}(\boldsymbol{x}) - \bar{y}(\boldsymbol{x})]^2\right]}_{\text{Bias}^2} + \underbrace{E_{\boldsymbol{x},y}\left[[\bar{y}(\boldsymbol{x}) - y]^2\right]}_{\text{Noise}}.$$

On the right-hand side, the first term is the *variance*. It depends on the particular choice of features used in the training data. It captures on the average how much the model predictions change if a different training set is used. If we have the best possible model, it measures how far we are from the expected model. The middle term on the right is the square of the bias. The *bias* is the inherent error of the model even with infinite training data. The term on the far right is noise. The *noise* is data intrinsic. For example, it might be due to a weak distribution of the data and choice of features. In general, you beat this by having better data.

When *underfitting* occurs, bias can cause the learning algorithm to miss relevant relations between features and labels. *Overfitting* can cause a high variance more indicative of the noise in

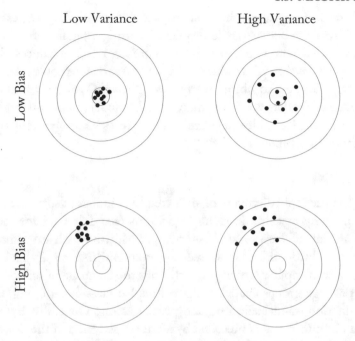

Figure 1.4: Schematic illustration of the meaning of high and low bias and high and low variance. each dot represents the results of one training set. After [28].

the training data than natural variations among the data. Empirically, one observes that models with low bias are usually complex. The standard example would be fitting data to a very high degree multi-variate polynomial. The higher the degree and the larger the number of independent variables the better the fit. We have come back to the dictum that with enough parameters you can fit anything. Ultimately, overfitting memorizes the data instead of learning it. Schematically, the situation is depicted in Figure 1.3. In the absence of noise, the variations of the contribution to the error from bias and variance are shown as a function of complexity. *Complexity* is a bit hard to define, but is some measure of the number of hyper parameters, the number of features and the ranges over which these parameters vary. The more of these quantities that are present the more complex is the model. Models with low bias do not necessarily have small error. In Figure 1.4, we schematically illustrate the meanings of high and low biases and variances. More fundamentally, reducing the dimensionally of the problem, for instance, by reducing the number of features and selecting better features tends to reduce variance. On the other side of the coin, adding features tends to decrease bias. Good cross-validation varies the complexity until a sweet-spot is found where the variation of the bias and variance combine to produce the smallest error. In some sense, we need to seek a Goldilocks solution: Too many and too few features are not right. Ultimately, for good model building we seek something in between that is just right.

Some machine learning algorithms, such as *LASSO* and *ridge* regression (Chapter 2), are designed to address the bias-variance problem by using a technique called regularization. Regularization adds a constraint to the optimization problem that introduces a bias but can often reduce the variance. *Ensemble methods* address the bias vs. variance problem by the techniques of bagging and boosting. *Bagging* combines many "strong" models, ones with low bias, to produce one with lower variance. *Boosting* combines many "weak" models, ones with high bias, to produce one with higher variance. Examples here are random forests and gradient tree boosting. We discuss these methods in Chapter 3.

1.3.3   TOOLBOXES

Browsing will lead to the rapid discovery of an incredible number of software packages available that encode one or many machine learning algorithms (see Table 1.1 for selected examples). We note the commercial software as Mathematica and Matlab which are commonly used in the physical sciences. In addition, there is the freeware *R* and *scikit-learn* [29]. Mathematica, Matlab, and *R* [30] are also computer languages. Python is the language of scikit-learn. The website for this software package groups available algorithms for classification, regression, clustering, dimensionality reduction, model selection and preprocessing. Generally, Python code is given for the use of each algorithm that is preceded by a brief description of the algorithm, often with references to the machine learning development literature. These packages also include software that greatly facilitates the cross-validation by such methods as a *grid-search* over ranges of the adjustable parameters and learning curves. The value of the examples is their serving as templates for intricate plotting of the results. Web browsing will also unveil a number of tutorials, lecture notes, and on-line courses devoted to the field. The website astroML.org, Machine Learning and Data Mining for Astronomy, contains tools, built upon scikit-learn and described in the text *Statistics, Data Mining, and Machine Learning for Astronomy* [9], that are sometimes useful for materials application. In the near future, we expect that materials science tools to be built upon scikit-learn or other software packages and downloadable from a materials-oriented web site. In general, the amount of software available from the Web for machine leaning is so vast that one should search before writing one's own.

1.4   NO FREE LUNCH THEOREMS

We now present an essay on a topic that is rarely mentioned in the materials machine learning literature, the *No Free Lunch Theorems* (NFLTs) [31–33]. They are a bit enigmatic. Their proofs are very abstract. Nevertheless, they say something about the limits of what we can expect to learn by inductive reasoning. They were developed when machine learning was experiencing its first growth spurt.

In the physical sciences, we typically look at data to establish general principles and then use these principles to predict beyond what we know. On the one hand, we are trying to use machine learning to help us extend our knowledge of materials beyond what we know. On the

Table 1.1: Selected examples of publicly available learning resources related to machine learning

Name	URL	Description
General-Purpose Machine-Learning Frameworks		
Deeplearning4j	https://deeplearning4j.org	Distributed deep learning for Java
Mlpack	https://mlpack.org	Scalable machine-learning in C++
R	https://www.r-project.org/	A language and environment for statistical computing and graphics
Scikit-learn	http://scikit-learn.org	Machine-learning and data-mining toolkits available from the scikit family of toolboxes
TensorFlow	https://www.tensorflow.org	A neural-network-based open-source platform for deep learning
Weka	https://cs.waikato.ac.nz/ml/weka	Collection of machine-learning algorithms and tasks written in Java
Tools Specific to Materials		
AFLOW ML	http://aflowlib.org/aflow-ml	Machine learning tools for materials property predictions
Amp	https://bitbucket.org/andrewpeterson/amp	Package to facilitate machine learning for atomistic calculations
ChemML	https://github.com/hachmannlab/chemml	An informatics program suite for chemical and materials data
DeepChem	https://deepchem.io	Python library for deep learning of chemical systems
DScribe	https://github.com/SINGROUP/dscribe	A Python package for creating machine learning descriptors for atomistic systems
MatMiner	https://hackingmaterials.github.io/matminer	A Python library for feature generation and data mining
NOMAD	https://analytics-toolkit.nomad-coe.eu	Collection of tools to explore correlations in materials datasets
Polymer Genome	https://www.polymergenome.org	An informatics platform for polymer property prediction and design using machine learning
PyMKS	http://pymks.org/	Informatics tools to solve multi-scale materials science problems

other hand, we can ask what the general principles are for using these methods. Unfortunately, there are at best a few. The bias vs. variance problem, while not a principle, is a general concern. The NFLTs are general and as theorems state irrefutable results. Indirectly, they say much of what is done in machine learning is heuristic. Experience justifies the methods and procedures.

While to date the toolbox perspective characterizes the use of machine learning methods in materials design and discovery, there is also the perspective of machine learning as an activity to develop those tools. At an even deeper level, there is a perspective asking what can we say about the underlying skeleton of such central topics as optimization theory and supervised learning before we flesh it out with particular contexts and problems. At this deeper level, the question becomes what we can say about the limits of inductive inference. This latter perspective leads us to the NFLTs [31–33]. They define what in general we can expect the machine learning algorithms to do and not do.

The most notable NFLTs are for search, optimization, and supervised learning. For optimization problems [31, 33–35], the theorem is often stated as "A universal optimizer does not exist." Sometimes, it is stated as "The only way one optimization strategy will outperform another is if it is specialized to the structure of a specific problem." It was originally stated as "Any two algorithms are equivalent when their performance is averaged uniformly over all possible problems."

These theorems are relevant to machine learning because most machine learning methods optimize something, typically performing a *constrained* fit of a specialized cost, loss, or utility function to the data. In general, the NFLTs for optimization imply that a learning algorithm which works well for one class of problems will work poorly for another. At least implicitly, the NFLTs seem to be a reason why for a given problem class there are so many methods. More practically, you might have several methods that you like more than others. For a given application, one will work better than the rest. Typically, you will need to try all to find out which one. If you change the problem, for example, by adding a lot more data, by adding and removing some features, etc., another of your favorite methods might rise to the top. A NFLT was in fact first introduced in the context of supervised learning [31–34].

As NFLTs developments and proofs take place in general contexts, they are quite abstract. To motivate many of their consequences, we adopt and summarize a simple heuristic due to Ho and co-workers [36–38]. This heuristic, as the theorems themselves, assume the world is discrete and finite. As machine learning is performed on a digital computers whose world is discrete and finite, this assumption does not seem to generate a significant loss of generality.

We start by assuming we have three sets: $X = \{x_0, x_1, \ldots, x_{|X|-1}\}$ where the size of X is $|X|$, $Y = \{y_0, y_1, \ldots, y_{|Y|-1}\}$ where the size of Y is $|Y|$ and $F = \{f_0, f_i, \ldots, f_{|F|-1}\}$ where the size of F is $|F| = |Y|^{|X|}$. The members of set F are the unique mappings $f(x) = y$ of the members of X into Y. In general, we do not know these mappings. These three sets represent the complete universe of the problem at hand. We now define a matrix F of $|X|$ rows and $|F|$

columns whose elements are $F_{ij} = f_j(x_i)$. For $|X| = 3$ and $|Y| = 2$, for example, we have

$$
F = \begin{array}{c|cccccccc}
 & f_0 & f_1 & f_2 & f_3 & f_4 & f_5 & f_6 & f_7 \\
\hline
x_0 & y_0 & y_1 & y_0 & y_1 & y_0 & y_1 & y_0 & y_1 \\
x_1 & y_0 & y_0 & y_1 & y_1 & y_0 & y_0 & y_1 & y_1 \\
x_2 & y_0 & y_0 & y_0 & y_0 & y_1 & y_1 & y_1 & y_1
\end{array}
\tag{1.3}
$$

Now we rewrite this matrix by mapping Y to the set of integers $\{0, 1, \ldots, |Y| - 1\}$

$$
F = \begin{array}{c|cccccccc}
 & f_0 & f_1 & f_2 & f_3 & f_4 & f_5 & f_6 & f_7 \\
\hline
x_0 & 0 & 1 & 0 & 1 & 0 & 1 & 0 & 1 \\
x_1 & 0 & 0 & 1 & 1 & 0 & 0 & 1 & 1 \\
x_2 & 0 & 0 & 0 & 0 & 1 & 1 & 1 & 1
\end{array}
\tag{1.4}
$$

The matrix is now of the form of a counting matrix about which we can prove several useful facts. The first is each y appears in each row $|Y|^{|X|-1}$ times. The second is the row sums and hence the row averages are all equal. The next one is that all the rows and all the columns are unique. Finally, if we pick any row i and any member y of Y, then the submatrix formed by eliminating row i and all columns j, that is, $F_{ij} \neq y$, is also a *counting matrix*. For example, if we eliminate row 0 and all columns such that $f_j(x_0) \neq y_0$ then the resulting matrix is

$$
F' = \begin{array}{c|cccc}
 & f_0 & f_2 & f_4 & f_6 \\
\hline
x_1 & y_0 & y_1 & y_0 & y_1 \\
x_2 & y_0 & y_0 & y_1 & y_1
\end{array}
$$

If we now eliminate x_2 and all columns for which $f(x_2) \neq y_0$, we get

$$
F'' = \begin{array}{c|cc}
 & f_0 & f_2 \\
\hline
x_1 & y_0 & y_1
\end{array}
$$

We are now in a position to discuss some NFLTs.

   If we let x be an optimization variable, f be a performance function, and $y = f(x)$ be the performance of x on f, what our F matrix says is when averaged over all possible performance functions, the average performance is independent of the row of F that is our choice of x. This is the essential idea of the NFLTs for optimization: No choice is universally better than another. If we were to give different interpretation to X and Y, using the properties of the resulting F matrix we would get other NFLTs. We note that this row average over all performance functions assumes each function contributes to the average equally; that is, we are assuming we have no *a priori* knowledge that one f should contribute more beneficially than another.

   Most generally, we can regard the rows of F as all possible strategies to solve problems, the columns as all possible problems, and an element of F as the performance of strategy x on problem f. The equality of the row sums means that no strategy outperforms all others on all problems. We can view this observation as a statement of the conservation of performance: If a given strategy performs well on some problems, it must perform poorly on all others.

If we look at optimization at bit more specifically, we would say we have a problem $\min_x f(x)$ where the x's are vectors of input parameters, f's are the performance (cost, utility, etc.) functions, and y's the values of the performance. We can solve this problem by searching X for the member that gives, say, the lowest cost. In a search algorithm for optimization, at the nth step our algorithm will have collected n distinct samples of X, observed their costs y and is trying to learn which f in F the algorithm is working on so to learn the optimum in the fewest samples possible. The collected x's are the history of the search.

The properties of F brings us back to a statement that no search algorithm should *a priori* expect to give better performance than another. We illustrate this in the following way: We take $X = \{x_0, x_1, x_2\}$ and $Y = \{0, 1\}$ where 0 is bad performance and 1 is good performance. We are back to our previous F matrix in (1.4). If we knew which column we were working on, then we could design an algorithm that would immediately find the optimal row. But since we assumed we had no prior knowledge about the columns, we can only sample X in trying to find the optimal one.

In our search, if $m = 1$, all columns are equally likely. No matter how the search algorithm picks the row the expected performance is the same because all rows averages are the same. Suppose we pick x_0 and find that $y_0 = f(x_0) = 0$. This means in principle we can shrink our F matrix by eliminating row 0 and all columns j such that $f_j(x_0) \neq 0$ to find

$$F' = \begin{array}{c|cccc} & f_1 & f_3 & f_3 & f_7 \\ \hline x_1 & 0 & 1 & 0 & 1 \\ x_2 & 0 & 0 & 1 & 1 \end{array}$$

The expected performance of each of the remaining rows is, however, the same. While the sampled x provided information about which columns we were not working on, they did not provides us with information about which ones we were working one. We gained no information to help guide the search. Thus, in the absence of prior assumptions about the cost functions being worked on, all search algorithms have the same expected performance.

We can gain further insights into optimization from the F matrix when it is expressed in terms of conservation principles. For each principle, we find there are trade-offs between what we want and what we do not want. First, we revisit conservation of performance: Average performance is conserved. We have motivated this statement under the assumption that each column was equally likely. What might change if we now assume we have some prior knowledge about the columns expressed as some probability $P(f)$? If we optimize expected performance by concentrating on high probability columns and our assumptions about the distribution of columns are wrong, then the performance of our optimal solution, in fact, could be arbitrarily bad. Even if our assumptions are correct, the solution for the best expected performance may occasionally allow very bad performance as long as it is offset by other performances that are generally only slightly better than average. Since each value of y appears an equal number of times in each row, there is no row that can assign a every bad outcome to a low probability column.

Next, we consider conservation of robustness: No strategy is universally more robust than another. A strategy is robust if it guarantees a certain level of performance over a large range of problems. From the properties of the counting matrix, since each y in Y appears in a row $|Y|^{|X|-1}$ times, all strategies give good performance for $|Y|^{-1}$ of the time; that is, no one is more robust than any other. On the other side of the coin, they all give bad performance with the same frequency. Every strategy is both robust and fragile in the same way.

If we now assume we have prior knowledge $P(f)$ about the importance of the columns, we might be able to increase robustness but doing so generally requires sacrificing some performance. To see this, we suppose our prior knowledge says to assign equal probabilities to the columns of a subset F_1 of those in F and let x_1 be the row that gives the optimal solution using this subset. Now we consider for a robust solution a second larger subset F_2 of F under the original assumption that columns are equally likely. Let x_2 be the robust solution. In general, x_1 and x_2 are not the same. If they are not the same, then the optimal performance of x_1 over the subset F_1 will be better than the robust performance of x_2 over the subset F_2. Thus, a robust solution must generally give up performance in return to sensitivity to errors in prior knowledge.

The easy to understand, just summarized heuristic of Ho and coworkers illustrates the essential implications of NFLTs. Things are easiest to understanding when each f is equally likely. When they are not, as hinted in the discussion of conservation of performance and robustness, and a distribution $P(f)$ is folded into the discussion, our statements seem less precise, but we are still faced with the fact that if we improve performance relative to some set of f's, it degrades the performance for the others. Cases where $P(f)$ is ingrained in the analysis were discussed by Wolpert and co-workers [31, 33, 34] well prior to the heuristic. They employed *Bayesian reasoning* (see Appendix A). We now simply state their main results first for optimization by searching and then for supervised learning.

For search algorithms for optimization, we assume we have collected a set of n data $\mathcal{D}_n = \{\mathcal{D}_x^n, \mathcal{D}_y^n\}$ of n paired values of x and y. With this data $\mathcal{D}_n$, we choose some algorithm A and build a performance measure $\Phi_{\mathcal{D}^n}$ that maps the $\mathcal{D}_y^n$ to real numbers. For a given x, the objective function f predicts y. The x and y eventually augment $\mathcal{D}_n$ to $\mathcal{D}_{n+1}$, and the process is repeated.

Wolpert and co-workers [31, 33–35] show that

$$P(\mathcal{D}_y^n | A, n) = \sum_f P(\mathcal{D}_y^n | f, A, n) P(f).$$

They note that in a discrete world the right-hand side is the form of an inner product of two vectors. One vector $P(f)$ is independent of the algorithm and the data, but it is something we otherwise do not know. The other vector $P(\mathcal{D}_y^n | f, A, n)$ expresses the probability of the performances given f, the algorithm, and the number of data. It gives the details about how the search algorithm works.

To derive an NFLT, they now average over a $P(f)$. To do this, they take B to be a subset of all the objective (cost) functions and use the above to prove that

$$\sum_{f \varepsilon B} E_{f,n,A}(\Phi) + \sum_{f \varepsilon F \backslash B} E_{f,n,A}(\Phi) = \text{constant}, \tag{1.5}$$

where the constant depends on Φ but is independent of A and B. The equation thus says that the expected performance $E_{f,n,A}(\Phi)$ is independent of the algorithm. The notation $F \backslash B$ means the set of cost functions F with the subset B removed.

The NFLT result was extended to the case of averages including an average over all $P(f)$. Here, because $P(f)$ is a finite, real vector normalized to unity, the possible $P(f)$'s live on a simplex Ω. Summations are over the elements π of the simplex Ω. If Π is a subset of elements of Ω, they show that

$$\int_{\pi \varepsilon \Pi} d\pi E_{\pi}(\Phi | n, A) + \int_{\pi \varepsilon \Omega \backslash \Pi} d\pi E_{\pi}(\Phi | n, A) = \text{constant},$$

where the constant depends on Φ but is independent of A and Π. This NFLT says that if the algorithm performs well for one set of $P(f)$'s in Π, it must perform poorly for those $P(f)$'s not in Π.

They also derive an inner product formula and an NFLT for supervised learning. The analysis is much more involved. In this case, the cross-validation training set of data $\mathcal{D}$ has a slightly different interpretation than $\mathcal{D}^n$ from the optimization case. It is not the set of values successively augmented by the search algorithm, that is, the history of the search, but is a set data that within the context of a single cross-validation step is fixed. What we want from this set is to learn the f so we can generalized beyond this set; that is, if we propose an x outside of $\mathcal{D}$, we want to state a y. We do this by applying a learning algorithm to the data $\mathcal{D}$ that represents a set of hypotheses $H = \{h_0, h_1, \ldots, \}$ with $h(x) = y$ labeling x. The learning algorithm is ultimately the conditional probability $P(h|\mathcal{D})$ which is learned and hence becomes known. What we want, however, is $P(f|\mathcal{D})$ which we do not know. The NFLT for optimization enters into the construction of the learning algorithm, which fixes the hyperparamters of a model defining the hypotheses. The hyperparameters define some set X' and its performance in learning the mapping h of some element of X' into Y. This is not $\mathcal{D}$ but another set of data $\mathcal{D}_n$ collecting the history of the learning. The NFLT enters through the selection of the h by the learning algorithm. With respect to the value of some cost c associated with the supervised learning, Wolpert [39] shows that

$$P(c|\mathcal{D}) = \int df dh P(f|\mathcal{D}) P(h|\mathcal{D}) M_{c,\mathcal{D}}(f, h),$$

where M is a matrix symmetric with respect to its arguments. This result is an inner product involving a known quantity $P(h|\mathcal{D})$ that expresses the algorithm and another $P(f|\mathcal{D})$ which is unknown. With some additional conditions, this inner product formula results in several NFLTs. In brief, in supervised learning, if some method does well for some problems, it must do poorly for the others.

The intent of the NFLTs seems to have been the generation of guidance for constructing tools. In many cases, their assumptions make them not directly applicable to practical problems. Their overarching interpretation is many machine learning procedures, such as cross-validation, are heuristic and lack formal justification. Whether these theorems were ever used in toolbox construction is questionable. What is likely is the experience gained by various attempts in the construction of similar tools for the same class of problems, provided the prior knowledge enables some approaches to become more favored over others. For a given problem class, we however cannot expect one approach to always be the best.

1.5   REFERENCES

[1] https://obamawhitehouse.archives.gov/mgi 1

[2] T. Lookman, F. J. Alexander, and K. Rajan, Eds. *Information Science for Materials Discovery and Design*, volume 225 of Springer Series in Materials Science. Springer International Publishing, Heidelberg, 2016. 3

[3] D. J. C. MacKay. *Information Theory, Inference and Learning Algorithms*. Cambridge University Press, Cambridge, 2003. DOI: 10.1108/03684920410534506 3

[4] C. M. Bishop. *Pattern Recognition and Machine Learning*. Springer-Verlag, New York, 2006.

[5] N. Christianni and J. Shawe-Tayor. *An Introduction to Support Vector Machines*. Cambridge University Press, Cambridge, 2000.

[6] C. E. Rasmussen and K. J. Williams. *Gaussian Processes for Machine Learning*. MIT Press, Cambridge, MA, 2006. DOI: 10.7551/mitpress/3206.001.0001

[7] T. Hastie, R. Tibshirani, and J. Friedman. *The Elements of Statistical Learning*. Springer, New York, 2008. 9

[8] P. Flach. *Machine Learning: The Art and Science of Algorithms that Make Sense of Data*. Cambridge University Press, New York, 2012. DOI: 10.1017/cbo9780511973000 7, 8, 10

[9] Z. Ivezić, A. J. Connolly, J. T. VanderPlas, and A. Gray. *Statistics, Data Mining and Machine Learning in Astronomy*. Princeton University Press, Princeton, NJ, 2014. DOI: 10.23943/princeton/9780691151687.001.0001 3, 8, 9, 10, 14

[10] K. T. Butler, D. W. Davies, H. Cartwright, O. Isayev, and A. Walsh. Machine learning for molecular and materials science. *Nature*, 559(7715):547, 2018. DOI: 10.1038/s41586-018-0337-2 3

[11] Y. Liu, T. Zhao, W. Ju, and S. Shi. Materials discovery and design using machine learning. *Journal of Materiomics*, 3(3):159, 2017. DOI: 10.1016/j.jmat.2017.08.002

[12] A. Jain, G. Hautier, S. P. Ong, and K. Persson. New opportunities for materials informatics: Resources and data mining techniques for uncovering hidden relationships. *Journal of Materials Research*, 31:977, 2016. DOI: 10.1557/jmr.2016.80

[13] M. A. Mosquera, B. Fu, K. L. Kohlstedt, G. C. Schatz, and M. A. Ratner. Wave functions, density functionals, and artificial intelligence for materials and energy research: Future prospects and challenges. *ACS Energy Letters*, 3(1):155, January 2018. DOI: 10.1021/acsenergylett.7b01058

[14] K. Takahashi and Y. Tanakab. Materials informatics: A journey towards material design and synthesis. *Dalton Transactions*, 45:10497, 2016. DOI: 10.1039/c6dt01501h

[15] Y. Lyu, Y. Lkiu, and B. Guo. Materials discovery and design using machine learning. *Journal of Materiomics*, 3:221, 2017. DOI: 10.1016/j.jmat.2017.08.002

[16] W. Lu, R. Xiao, J. Yang, H. Li, and W. Zhang. Data mining-aided materials discovery and optimization. *Journal of Materiomics*, 3:191, 2017. DOI: 10.1016/j.jmat.2017.08.003

[17] X. Zhang and Y. Xiang. Combinatorial approaches for high-throughput characterization of mechanical properties. *Journal of Materiomics*, 3:209, 2017. DOI: 10.1016/j.jmat.2017.07.002

[18] S. Curtarolo, G. L. Hart, M. B. Nardelli, N. Mingoand, S. Sanvito, and O. Levy. The high-throughput highway to computational materials design. *Nature Materials*, 12:191, 2013. DOI: 10.1038/nmat3568

[19] J. M. Rondinelli, N. A. Benedek, D. E. Freredman, A. Kovner, E. E. Rodriguez, E. S. Toberer, and L. W. Martin. Accelerating functional materials discovery: Insights from geological sciences, data-driven approaches, and computational advances. *American Ceramics Society Bulletin*, 92:14, 2013.

[20] T. Mueller, A. G. Kusne, and R. Ramprasad. Machine learning in materials science: Recent progress and emerging applications. *Reviews in Computational Chemistry*, 29:186, 2016. DOI: 10.1002/9781119148739.ch4 3

[21] J. R. Chelikowsky and J. C. Phillips. Quantum-defect theory of heats of formation and structural transition energies of liquid and solid simple metal alloys and compounds. *Physical Review B*, 17:2453, 1978. DOI: 10.1103/physrevb.17.2453 3

[22] E. Mooser and W. B. Pearson. On the crystal chemistry of normal valence compounds. *Acta Crystallographica*, 12:1015, 1959. DOI: 10.1107/s0365110x59002857 3

[23] J. St. John and A. N. Block. Quantum-defect electronegativity scale for nontransition elements. *Physical Review Letters*, 33:1095, 1974. DOI: 10.1103/physrevlett.33.1095 4

[24] A. Zunger. Systematization of the stable crystal structure of all AB-type binary compounds: A pseudopotential orbital-radii approach. *Physical Review B*, 22:5839, 1980. DOI: 10.1103/physrevb.22.5839 5

[25] D. Pettifor. Phenomenological and microscopic theories of structural stability. *Journal of Less Common Metals*, 114:7, 1985. DOI: 10.1016/0022-5088(85)90384-4 5

[26] G. Pilania, J. E. Gubernatis, and T. Lookman. Structure classification and melting temperature prediction in octet AB. *Physical Review B*, 91:214302, 2015. DOI: 10.1103/physrevb.91.214302 5

[27] M. Ashby. *Material Selection in Mechanical Design*. Butterworth-Heinemann, Burlingham, 2008. DOI: 10.1016/b0-08-043152-6/00910-4 5

[28] http://scott.fortmann-roe.com/docs/BiasVariance.html 12, 13

[29] F. Pedregosa, G. Varoquaux, A. Gramfort, V. Michel, B. Thirion, O. Grisel, M. Blondel, P. Prettenhofer, R. Weiss, V. Dubourg, J. Vanderplas, A. Passos, D. Cournapeau, M. Brucher, M. Perrot, and E. Duchesnay. Scikit-learn: Machine learning in Python. *Journal of Machine Learning Research*, 12:2825, 2011. 14

[30] R Developmewnt Core Team. *R: A Language and Environment for Statistical Computing*. R Foundation for Statistical Computing, Vienna, Austria, 2013. 14

[31] D. H. Wolpert. The lack of a proir distribution between learning algorithms and the existence of a priori distinctions between learning algorithms. *Neural Computation*, 8:1341, 1996. DOI: 10.1162/neco.1996.8.7.1341 14, 16, 19

[32] D. H. Wolpert. The lack of a prior distribution between machine learning algorithms and the existence of a priori distinctions between learning algorithms. *Neural Computation*, 8:1391, 1996. DOI: 10.1162/neco.1996.8.7.1341

[33] D. H. Wolpert and W. G. Macready. No free lunch theorems for optimization. *IEEE Transactions on Evolutionary Computation*, 1:67, 1997. DOI: 10.1109/4235.585893 14, 16, 19

[34] D. H. Wolpert. The supervised learning no-free-lunch. In R. Roy, M. Koppen, S. Ovaska, T. Furuhashi, and F. Hoffman, Eds., *Soft Computing and Industry*, p. 25, Springer-Verlag, London, 2002. 16, 19
DOI: 10.1007/978-1-4471-0123-9_3

[35] D. H. Wolpert. What the no free lunch theorems really mean: How to improve search algorithms. *Technical Report*, Santa Fe Institute, Santa Fe, NM, 2012. 16, 19

[36] Y.-C. Ho and D. L. Payne. Simple explanation of the no free lunch theorem of optimization. *40th IEEE Conference on Decision and Control*, Orlando, FL, December 2001. DOI: 10.1109/CDC.2001.980896 16

[37] Y.-C. Ho and D. L. Payne. Simple explanation of the no free lunch theorem and its implications. *Journal of Optimization Theory and Applications*, 115:549, 2002. DOI: 10.1023/a:1021251113462

[38] Y.-C. Ho, Q.-C. Zhao, and D. L. Payne. The no free lunch theorems: Complexity and security. *IEEE Transactions on Automatic Control*, 48:783, 2003. DOI: 10.1109/tac.2003.811254 16

[39] D. H. Wolpert. On the connection between in-sample testing and generalization error. *Complex Systems*, 6:47, 1992. 20

CHAPTER 2

Materials Representations

In this chapter, we discuss perhaps the most important component of machine learning model building—the one that converts a molecular or material system into a numerical representation. To build the models, we use a machine learning algorithm. What we will be discussing is defining the random variables on which the models depend. The chosen statistical learning algorithm uses this material representation to either quantify (dis)similarity between any given pair of materials across a target chemical space or extract useful and generalizable chemical insights from available data [1]. The choice of a suitable representation or *fingerprint* is problem-specific and largely depends on mechanistic details of the material phenomena being addressed [3–5]. Therefore, the identification of a relevant representation often requires significant domain expertise to ensure that the fingerprinting encodes all essential and easily accessible attributes of the target materials (or molecular) class. These material variables are distinct from what are called the hyper-parameters of the models. The learning algorithm adjusts the hyper-parameters to fit the data as a function of the random variables. The numerical representation is also called a descriptor or feature vector. In what follows, we use these terms synonymously.

2.1 CONDITIONS FOR A VALID REPRESENTATION

The process of selecting a material's representation provides an opportunity to integrate problem-relevant physics into the model and largely governs the model's generalizability, causality, and predictive power [6, 7]. The selection of an appropriately resolved numerical representation is dictated by the problem under study and the accuracy requirements of the predictions. In addition to application specific requirements, any viable materials representation also needs to satisfy a set of general requirements. Since within a statistical learning framework a representation serves as a surrogate (proxy) for a real material system, it has to be *unique*; that is, no two different materials should be represented by the same descriptor—the mapping from the materials to the representations should be one-to-one. It is also highly desirable that the representation be *invariant* to transformations preserving the target property of the system. For instance, transformations such as translations, rigid rotations, and permutations of like element species for materials are not expected to alter any of its properties. A non-trivial example is a rock salt ordered bulk binary lattice of A and A' elemental species, where simply interchanging the two sublattices (that is, swapping A and A') should leave the bulk solid unaltered. An effective representation must have such invariances built-in as learning these trivial properties from data can be highly inefficient and resource intensive, and might also result in poor predictive accuracy. Further, *easily accessible* attributes are preferred to ensure overall efficiency of the machine learning based surrogate model building exercise. For certain specific applications, such as learning local forces and potential energy surfaces, a *continuous* representation (in the configurational feature space) is preferred, as discontinuities work against the smoothness assumption of most machine learning models. Finally, we note that it also helps if the adopted fingerprinting scheme is *general* in the sense of being able to handle any combination of compositional and configurational complexity for finite molecular or periodic materials systems on an equal footing.

2.2   HIERARCHY OF MATERIALS REPRESENTATIONS

Specific choice and details of a suitable representation depend on the problem being addressed and the desired accuracy of the predictions. Depending on these factors, one can design representations with varying levels of granularity. As an example, if we are only interested in gaining a high-level qualitative understanding of some selected materials properties, such as catalytic activity, electronic properties, or mechanical strength, across a well-defined chemical space and the quantitative accuracy of the predictions is not critical, then it might be sufficient to define the representation at a coarse level. Such a representation would typically include, for example, relevant bulk properties of compounds (bandgaps, density, elastic constants), key properties of elemental species forming the compound (electronegativities and ionization potentials of isolated atoms), or even gross-level structural features (grain size, dislocation density). Such *macroscopic representations* are very effective in developing physically meaningful cheaper surrogate models of complex materials phenomena and in enabling automated extraction of chemical insights and design rules for novel materials discovery while sweeping across a large set of diverse chemistries [8–16].

   On the other hand, if the aim is to predict subtle changes in total energies or accurate predictions of local forces on atoms while exploring potential energy surfaces of certain molecules or solids, a representation that encodes details of the local atomic configuration with a sufficiently high resolution to meet the desired chemical accuracy in the predictions is needed. Such fine representations that encode details of atomic-level structural information reaching a sub-Angstrom scale in resolution, henceforth referred to as *microscopic representations*, are frequently employed to develop advanced machine learning determined force fields for finite temperature molecular dynamics simulations of large scale systems (that is, system sizes beyond the reach of standard first principles-based quantum mechanical simulations). We note that the general uniqueness and invariance conditions discussed above are quite critical for the case of microscopic representations and need careful consideration for developing an effective representation. Doing this is currently a highly active area of research [17, 19–29].

   While macroscopic and microscopic representations focus largely on capturing compositional and configurational details, respectively, of the material system under study, for certain applications it becomes necessary to account for both the chemistry and some level of detail about the local structure to be able to quantitatively describe target materials properties. For instance, certain properties of a polymer system are described in terms of different building blocks and motifs that form the polymer chain. In this case, while consideration of atomic arrangements within each motif might not be critical, fractional compositions of the building blocks and how different motifs are arranged within a given polymer dictates many physically measurable properties of the system. This class of representations where local configuration information appears only in coarse-grained manner are referred to as *mesoscopic representations*. This class of fingerprints—finding their conceptual roots in well-established classic approaches such as the group contribution approach [30], cluster expansion approach [31, 32], or quantitative structure property relationships (QSPR) [33]—provide an effective statistical learning pathway to study chemical trends in diverse materials systems spanning both composition and configurational spaces [34–38].

Figure 2.1 illustrates materials representations falling within the three different categories of the adopted hierarchy. Although the above discussion developed a convenient classification scheme to categorize materials representations at varying levels of granularity, keywords such as *microscopic* or *macroscopic* used in the descriptor classification are only for a relative comparison and do not refer to the relevant length scales. In fact, the resolution of the microscopic representations would be invariably at a sub-Angstrom level. Further, it is always possible (and can be practically useful) to combine representations at different scales to tune the underlying cost-accuracy tradeoff. As a general rule, the finer-resolved the fingerprint, the greater the expected accuracy (and higher the computational cost) of the learning framework. Lastly, we point out that despite all these considerations pertaining to domain knowledge, the nature of the problem being addressed, dimensionality and complexity of the materials chemical space, target performance metrics, availability of data, etc., construction of an *optimal* fingerprint is never straightforward and should be pursued with care. There is also debate about which feature selection strategy is the best: Whether a fingerprint should be selected *a priori* solely based on expert opinion or whether it should be *mined* automatically from the training data after considering a large number of possibilities remains an interesting open question in the community. Nevertheless, many exciting developments in fingerprinting materials and molecular systems are occurring. In what follows, we discuss various representations in further detail, highlighting selected examples from recent literature.

2.2.1   MICROSCOPIC REPRESENTATIONS

To start, we focus on representations that allow us to "code" materials or molecules at the finest possible scale and enable us to make detailed comparisons between any two closely related configurations. Naturally, any such representation must incorporate a sub-Angstrom-level of detail of the local atomic structure and exhibit sensitivity to the subtle changes in the relative arrangements of atoms. Consequently, microscopic representations are usually the most expensive to compute and lead to the most accurate predictions when used in a machine learning framework.

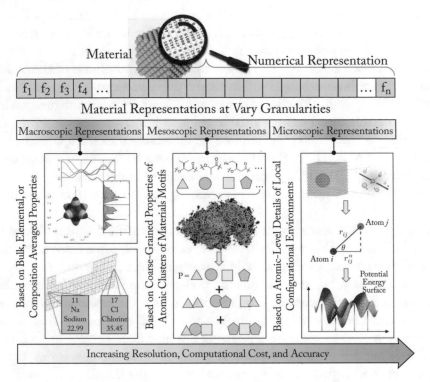

Figure 2.1: A graphical depiction of hierarchical granularity available for materials representations often employed for statistical learning problems in materials science and related fields. The macroscopic or gross-level representations are either formed by easily accessible bulk properties of the compounds or by a selected set of elemental properties of the atomic constituents forming the compounds. The mesoscopic or the intermediate level descriptors utilize a coarse-grained description for a pre-specified set of motifs and how these motifs are combined together to form different compounds. Lastly, the microscopic or atomically resolved representations are formed by encoding local configurational details.

As one can imagine, such numerical representations are useful in diverse scenarios. Most commonly, they are employed when the target configurational space is vast and one is interested in learning a mapping from a compositional or configurational space to the corresponding material property space.

In the past several years, microscopic representations have been employed to create surrogate models that provide alternative pathways to outputs of quantum mechanical computations, with a particular emphasis on DFT-based computations—the current workhorse for solving the electronic structure problems [39, 40]. At the most fundamental level, bypassing the Kohn-Sham equation [40] in a quest to learn the local charge density and the corresponding electronic

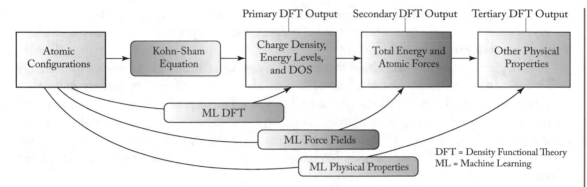

Figure 2.2: The hierarchical-paradigm of learning outputs of DFT computational (or quantum mechanical computations, in general) via machine learning-based surrogate models. The primary, secondary, and tertiary outputs refer to the electronic charge density and energy levels, total energy and local atomic forces, and all other physical properties of interest, respectively. Adapted from [29], with permissions.

energy levels directly given just the positions and identities of the atoms is the aim. The *primary* outputs from such calculations can then be used to compute *secondary* outputs such as the total potential energy of the system and atomic forces. Subsequently, other derived physical properties can be computed as *tertiary* outputs. Alternatively, surrogate machine learning models can be constructed to predict directly the secondary outputs, that is, total energies (using global representations) or atomic forces (using local representations). It is also possible to construct machine learning models for predicting tertiary outputs directly. This hierarchical-paradigm of applying microscopic representations-based surrogate models to different outputs of first-principles calculations is schematically captured in Figure 2.2.

Toward learning the primary outputs of the Kohn-Sham equation (or the Schrödinger equation in general) by using a statistical learning framework, recent endeavors have looked promising [29, 41–46]. While most efforts have focused on learning specific terms in the Kohn-Sham Hamiltonian, such as the exchange correlation and kinetic energy terms, a more recent breakthrough has reported a successful mapping of the charge density and the entire local density of states (LDOS) spectrum to the local atomic environment with an unprecedented predictive accuracy. In this study, Chandrasekaran et al. [29] utilized a new rotationally invariant microscopic representation to map the atomic environment around points on a pre-specified fine grid to the electron density and local density of states at that grid-point. They learned this mapping by using a neural network trained on previously generated reference DFT results at millions of grid-points and subsequently used the mapping for predicting the electronic structure of aluminum metal and polyethylene insulator with a very high fidelity. The ability to successfully learn charge density and LDOS models opens transformative pathways that allow accurate electronic structure predictions on large ensembles of atoms, spanning time and length scales beyond the reach of state-of-the-art DFT computations. Moreover, following this approach, one can efficiently determine the aforementioned secondary and tertiary outputs of DFT from the machine learning based predictions.

A larger number of studies have focused on using surrogate models to predict total energy and atomic forces in molecules and materials, i.e., the aforementioned secondary outputs of quantum mechanical computations. For accurate and efficient predictions, these efforts have relied on a variety of carefully devised microscopic representations such as the many-body tensor representation [47], the SOAP descriptor [48], the Coulomb matrix representation [17], the Behler-Parrinello symmetry functions [49], and others [50–52]. Some of these are discussed below.

A much larger emphasis on predictions of total energy and local atomic forces is not surprising since there is great value in this enterprise as the developed predictive models provide effective alternative strategies with improved cost-accuracy tradeoffs for molecular and materials modeling. Present materials modeling approaches fall into two broad categories which differ in the accuracy (and the concomitant computational cost) of how well they predict interatomic forces or total energies. Quantum mechanics-based methods, including DFT, are versatile and offer the capability to accurately model a range of chemistries and chemical environments [53]. However, these methods remain computationally very demanding, limiting both the length and time scales of phenomena (to ~nanometers and ~picoseconds, respectively) [54]. Semi-empirical methods capture the essence of the interatomic interactions in a coarse-grained manner (via parameterized analytical functional forms) and are thus an inexpensive solution to the materials simulation problem [55–58]. Nevertheless, their applicability is severely restricted to the specific chemistries and chemical environments considered during parameterization, and accuracy cannot be guaranteed for properties not explicitly targeted by the fit. The machine learning-based approach has a potential to fill this gap by enabling high fidelity predictions of the secondary outputs (that is, the total energies and local forces) at relatively low computational cost—comparable or lower than semi-empirical methods. This alternative route to total potential energies and atomic forces has led to force fields and improved potentials [48–50, 52] which show promise for overcoming several limitations encountered by both classical and quantum molecular dynamics simulations.

Lastly, the ability to directly and rapidly predict a set of pre-specified physical properties (that is, the tertiary output of quantum mechanical computations) over a vast chemical space is useful for accelerated materials discovery [38, 59, 60]. Most, if not all, materials design and discovery problems are multi-objective in nature (discussed further in Chapter 5), and experimental measurements or first principles computations of all of the constraints is often prohibitively time and resource intensive. Surrogate models for physical properties of molecular and periodic systems show great promise for exploring multi-dimensional property spaces in search of target functionalities, as well as for revealing hidden correlations among a set of properties.

Owing to the wide range of applications and utility, a lot of effort is being devoted to the development of effective microscopic materials representations. As a result, a number of innovative schemes to numerically encode compositional and configurational details of molecules and materials at a finest level of granularity have been put forward. These schemes roughly divide into two groups. Those in the first group focus on encoding global representations to predict total energy directly to build surrogate models of potential energy surfaces. Those in the second group, on the other hand, take a local perspective, focus on atomic forces as a primary target property, and assume that the force on a given atom is completely determined by its local chemical environment. The total energy is then made accessible via a path integration using Newton's equations of motion. There are also representations designed to be general and can fit in either group. More recently, approaches are being developed to explore potential energy surfaces by combining both energies and force predictions. While an extensive review of the entire landscape of microscopic representations is impossible here, below we briefly provide a flavor for some of the selected ones.

Atom-Centered Symmetry Functions and Related Representations

To model high-dimensional potential-energy surfaces, in 2007 Behler and Parrinello [49] produced an elegant representation in terms of atom-centered symmetry functions (ACSF). Within this class of representations, the chemical environments of the atoms are characterized by a set of many-body mathematical functions, which depend on the positions of the neighboring atoms up to a cutoff radius R_c. The locality of the atomic interactions follows from the nearsightedness principle in quantum chemistry. The most prominent ACSFs are the "radial function" (G_i^R) and the "angular function" (G_i^A) represented as

$$G_i^R = \sum_{j=1}^{N_{atom}} e^{-\eta(R_{ij}-R_s)^2} f_c(R_{ij}), \tag{2.1}$$

$$G_i^A = 2^{1-\zeta} \sum_{j,k\neq i}^{all} (1-\cos\theta_{ijk})^\zeta e^{-\eta(R_{ij}^2+R_{jk}^2+R_{ik}^2)} f_c(R_{ij}) f_c(R_{jk}) f_c(R_{ik}). \tag{2.2}$$

In Equation (2.1), the radial function is a sum of Gaussians of atomic pairwise distances R_{ij}, and R_s is a parameter that shifts the positions of these Gaussians. The width of the Gaussians is controlled by η, and typically a number of different η values is used to obtain a radial fingerprint of the chemical environment. The angular ACSF given by Equation (2.2) represents an angular fingerprint of the environment using the angles θ_{ijk} formed by the ith atom with each pair of neighboring atoms j and k. The angular resolution is controlled by the parameter ζ, while $\lambda = \pm 1$ defines the positions of the extrema of the cosine function. The f_c is a cutoff function that ensures a smooth decay of the ACSFs with a slope of zero at a pre-specified cutoff radius

R_c. It is given by

$$f_c(R_{ij}) = \begin{cases} \frac{1}{2} \cdot [\cos(\frac{\pi R_{ij}}{R_c}) + 1], & \text{for } R_{ij} \leq R_c \\ 0, & \text{for } R_{ij} > R_c \end{cases}. \tag{2.3}$$

It is typical to use several tens of symmetry functions per atom with different specific values of the parameters η, ζ, R_s, and λ, while the cutoff radii R_c are determined as part of a convergence analysis ensuring that all energetically relevant interactions are included. We note that all ACSFs by construction present a rotationally and translationally invariant description of the local chemical environment because they depend on the internal radial and angular coordinates alone. The ACSFs are being successfully employed to construct a number of potentials, where the microscopic representations are used to map the local chemical environment onto the local atomic energies E_i and the total energy of the entire system is then determined by simply adding all the local energy contributions of the individual atoms. We further note that the summation over individual atoms is expressible in an alternative form using pairwise interaction contributions E_{ij}:

$$E = \sum_{i=1}^{N_{atom}} E_i = \sum_{i=1}^{N_{atom}} \sum_{j>i}^{N_{atom}} E_{ij}. \tag{2.4}$$

Analogous to the form of the ACSFs, the above is being been used to construct potentials in terms of pair-centered symmetry functions, which are equally suitable for obtaining high-fidelity description of potential energy surfaces.

In the past two decades, the ACSF-based descriptors have been extensively used to construct a variety of neural network based potentials for diverse systems such as small molecules, relatively simple extended systems (that is, model materials whose complexity has been reduced by freezing a majority of the degrees of freedom) such as small molecules interacting with frozen metal surfaces or complex systems such as water with the polarization effects included [61–65]. Further details on applications of the ACSF-based neural network potentials are found in recent reviews [23, 61, 66].

While the Behler approach largely focused on construction of potentials for total energy predictions, a closely related but suitably modified version of the radial ACSFs was recently proposed for the prediction of local atomic forces. Since the local force experienced by an atom placed in a certain chemical environment is a vector quantity, the radial ACSFs representation, in addition to being invariant to translations and rotations of the system, and to permutations of like atoms, must also be directionally resolved and proportionately change with small changes of the atomic arrangement. This is accomplished by defining a d-dimensional vector $\mathbf{V}_{i,\alpha}$ representing the local atomic environment of atom i viewed along the Cartesian direction α [24, 50]. For elemental materials, the sth component ($s \in \{1, d\}$) of this vector is defined as [49, 50, 67]:

$$V_{i,\alpha;s} = \frac{1}{\sqrt{2\pi}w} \sum_{j \neq i}^{N_{atom}} \frac{R_{ij}^\alpha}{R_{ij}} e^{-\frac{(R_{ij}-R_s)^2}{2w^2}} f_c(R_{ij}). \tag{2.5}$$

This representation is closely related to Equation (2.1), except for the term $\frac{R_{ij}^\alpha}{R_{ij}}$, where R_{ij}^α is the component or projection of R_{ij} onto the α direction and the term itself characterizes the contribution of atom j to the α component of the force on atom i. The pre-factor $\frac{1}{\sqrt{2\pi}w}$ is a normalization constant with the parameter w related to the width of the Gaussian. The above representation was employed to develop force fields for a variety of elemental solids, including Al, Cu, Ti, W, Si, and C, in a variety of different crystal structures [24, 50, 67]. A generalization of this representation was recently predicted electronic charge density and LDOS of metallic and insulating systems with an excellent performance [29].

Coulomb Matrix-Based Representations

A simple, effective, and physically intuitive representation, known as the Coulomb Matrix representation, was first employed by Rupp et al. to model molecular atomization energies [17]. Within this representation, a molecule is expressed as a symmetric matrix whose elements are given by:

$$X_{ij} = \begin{cases} 0.5 Z_i^{2.4} & \text{for } i = j \\ Z_i Z_j \Phi(\mathbf{r_i}, \mathbf{r_j}) & \text{for } i \neq j, \end{cases} \tag{2.6}$$

where $\Phi(\mathbf{r_i}, \mathbf{r_j}) = 1/||\mathbf{r_i} - \mathbf{r_j}||_2$ is the bare Coulomb potential with $|| \cdot ||_2$ denoting the Euclidian norm and Z_i and $\mathbf{r_i}$ represent the atomic number position of the ith atom in a molecule. The non-diagonal elements in this representation correspond to the pair-wise Coulomb repulsion between the positive atomic cores, and the diagonal elements are found from a polynomial fit of the nuclear charges to the total energies of the free atoms. This representation essentially uses the same input as electronic-structure calculations, namely nuclear charges and atomic positions, to learn molecular energies. A schematic representation of Coulomb Matrix for ethanol molecule (Figure 2.3a) is shown in Figure 2.3b, where the nature of the diagonal on-site and off-diagonal pair terms is qualitatively identified by using the atomic species that are responsible for the specific terms. In practice, unique elements of the representation (that is, diagonal elements and either the upper or lower triangle of the off-diagonal elements) are flattened into a vector to represent a molecule in a statistical learning exercise. Surrogate models built to predict total energies of molecules based on this representation have shown significantly higher accuracies than standard semi-empirical methods such as PM6 or simple bond-counting schemes [68, 69]. The Coulomb matrix approach has been extended to solids by using simple periodic concepts [20].

By using different forms of $\Phi(\mathbf{r_i}, \mathbf{r_j})$, in particular higher values of exponent n in the expression $1/|\mathbf{r_i} - \mathbf{r_j}|^n$ [70], a slight improvement over the original Coulomb matrix can be achieved

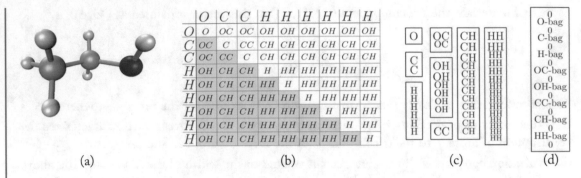

	O	C	C	H	H	H	H	H	H
O	O	OC	OC	OH	OH	OH	OH	OH	OH
C	OC	C	CC	CH	CH	CH	CH	CH	CH
C	OC	CC	C	CH	CH	CH	CH	CH	CH
H	OH	CH	CH	H	HH	HH	HH	HH	HH
H	OH	CH	CH	HH	H	HH	HH	HH	HH
H	OH	CH	CH	HH	HH	H	HH	HH	HH
H	OH	CH	CH	HH	HH	HH	H	HH	HH
H	OH	CH	CH	HH	HH	HH	HH	H	HH
H	OH	CH	CH	HH	HH	HH	HH	HH	H

(a)   (b)   (c)   (d)

Figure 2.3: Schematic illustration of the Coulomb Matrix and Bag of Bonds (BoB) representations. (a) A ball-and-stick representation of ethanol (CH_3CH_2OH) molecule and (b) involved onsite (diagonal) and pair (off-diagonal) interactions representing each Coulomb matrix element. (c) Different Coulomb matrix entries for ethanol are subsequently sorted into bags, leading to the BoB vector (d) obtained by concatenating these bags and adding zeros to allow for dealing with other molecules with larger bags. After [22], reproduced with permission.

with the best performance occurring for $n = 6$, a value reminiscent of London dispersion interactions [2]. As a result, the representation is also called the London matrix, and the improved performance has been rationalized in terms of an improved trade-off between the description of more localized covalent bonding and long-range intramolecular non-covalent interactions as compared to the expression used in the original Coulomb matrix.

It is straightforward to verify that the Coulomb Matrix representation, based solely on internal coordinates, is invariant with respect to rotations, translations, and symmetry operations such as mirror reflections or permutations of like atoms of a molecule in a three-dimensional space. However, there are several problems associated with this representation that make it difficult to represent molecules or materials in a vector space [19]. The first is that different numbers of atoms in different molecules result in different dimensionalities of the Coulomb matrices, which can complicate further mathematical treatment since a machine learning framework usually requires feature vectors with a fixed dimension to represent different molecules. This problem is easily addressed by introducing non-interacting "dummy atoms" with zero nuclear charge with the other atoms. In the Coulomb matrix representation, this introduction is achieved by padding each matrix with additional zeros, such that all molecular representations in the dataset have matrices of size $N \times N$ where N is the number of atoms in the molecule containing the largest number of atoms. The second problem, which turns out to be more difficult to address, is that there is no well-defined ordering of the atoms in the Coulomb matrix. For a given molecule with N number of atoms, one can obtain $N!$ different Coulomb matrices which are related with each other by simultaneous permutation of certain rows and columns. To overcome this non-uniqueness, a number of different approaches have been adopted. One such approach resorts to sorted eigenvalues of the original Coulomb matrix to represent each molecule as a vector. While this representation is unique and invariant with respect to permutations of the rows and columns of the Coulomb matrix, it leads to a drastic reduction in the dimensionality which in turn causes a loss of information and the introduction of unfavorable noise. An alternative approach focuses on a sorted Coulomb matrix, where a specific pre-specified prescription is used to sort a rows and columns of a computed Coulomb matrix. One way to do this is to permute the matrix in such a way that the rows (and columns) of the Coulomb matrix are ordered by their norm. This ensures a unique Coulomb matrix representation; however, as a downside, small variations in atomic coordinates sometimes cause changes in the Coulomb matrix, which might change the ordering abruptly. This disruption introduces unfavorable roughness in feature space which severely hinders the learning of structural similarities.

Bag of Bonds and Other Many Body Representations

Soon after the applications of Coulomb matrix based representations for molecular systems, it was realized that collective effects beyond pairwise potentials are important for chemically accurate modeling of molecular properties, and therefore these effects should be included in an effective representation that goes beyond the Coulomb matrix. To include these effects, a Bag of Bonds (BoB) descriptor was proposed based on concepts directly inspired by text mining bag-of-words descriptors frequently utilized in natural language processing in computer science to account for the frequency of occurrence of specific words in a given text to solve a classification problem such as "spam" or "no-spam" classification [22]. In a close analogy, the BoB descriptor relies on use of inverse interatomic distances for accurate predictions throughout chemical compound space.

As shown schematically in Figs. 2.3c and 2.3d, the BoB descriptor maps the molecular Hamiltonian onto a vector composed of bags, where each bag is labeled with a particular bond type (such as, C—C, C—N, O—H, etc.). Naturally building upon the Coulomb matrix concept, each entry in every bag is computed as $Z_i Z_j / |\mathbf{r_i} - \mathbf{r_j}|$, that is, as the off-diagonal elements of the Coulomb matrix, where $\mathbf{r_i}$ and $\mathbf{r_j}$ are the position vectors of the two atoms participating in a given bond. Next, this information is vectorized by simply concatenating all BoBs in a pre-specified order and padding each bag with sufficient zeros to give the bags equal sizes across the set of targeted molecular chemical space. A Gaussian or Laplacian kernel [18] built on a a suitably defined pairwise chemical distance between two molecules is then employed in subsequent learning process.

To understand the implicit many body effects buried in the BoB descriptor, we can decompose the BoB Laplacian kernel for two molecules M and M_I (with the index I labeling the training set) as $\exp(-\sum_j^n |M^j - M_I^j|/\sigma) = \Pi_j^n \exp(-|M^j - M_I^j|/\sigma)$. The index j runs over all n bags of pre-specified bonds, and as a result, both the Taylor-series expansion of the exponential (defined in terms of internuclear Coulomb repulsion) and the subsequent product includes contributions up to infinite order in terms of bond pairs between molecules M and M_I.

This reasoning argues that the BoB descriptor incorporates essentially the same ingredients as conventional many body potentials, albeit with a different and more general functional form. Naturally, the BoB representation is also invariant under molecular rotations and translations, while the permutational invariance is enforced by appending different bags in a pre-specified fixed order.

While many body effects are implicitly encoded in BoB representation, a recently reported generalization of this approach, referred to as the many-body tensor representation (MBTR), accounts for these effects directly by including bond, angle, and dihedral terms explicitly into the representation. Furthermore, while most representations are designed to encode either atoms in their chemical environment (symmetry functions) or molecular systems as a whole (Coulomb matrix, BoB), the general implementation of MBTR provides enough flexibility to switch between the local and global representations on demand, with an ability to treat molecular and periodic solids on an equal footing [47].

The Smooth Overlap of Atomic Positions (SOAP) Approach

Another systematic approach to describe the atomic environments for construction of machine learning potentials for high-dimensional systems has been put forward by Csányi and coworkers [48, 71]. The starting point for a representation within this so-called *bispectrum of the neighbor density* approach is the construction of a neighbor density $\rho_i(\mathbf{R})$, where $\mathbf{R}$ denotes the position vector in the space of each atom i. Practically, this is achieved by starting from a reference atom and placing δ functions at all positions of neighboring atoms in the environment up to the cutoff f_c (given by the same expression used in Equation (2.3)) as well as the position of the reference atom:

$$\rho_i(\mathbf{R}) = \delta(\mathbf{R}) + \sum f_c(R_{ij}) w_j \delta(\mathbf{R} - R_{ij}). \tag{2.7}$$

While the cutoff function f_c ensures a smooth decay to zero at a pre-specified radial distance R_c, the dimensionless weights w_j allow for a discrimination of different elemental species. The angular distribution of the neighbors is then obtained in a subsequent step that projects all the neighbors onto a sphere to give a projected density, which is eventually expanded in spherical harmonics.

Based on the assumption that the bispectrum components are linearly related to the atomic energies, a number of machine learning -based potentials, such as spectral neighbor analysis potentials (SNAP) [52] and Gaussian approximation potentials (GAP) [48], have been introduced. However, an obvious drawback of the above representation stems from its use of δ-functions in the neighbor density defined in Equation (2.7). Due to the lack of smoothness in the representation, slight deviations of the positions of atoms between two closely related environments can lead to strong numerical changes. This issue was addressed by proposing a new representation, known as the *smooth overlap of atomic positions* (SOAP) descriptor [72], where the δ-functions are replaced by smoother Gaussian counterparts, centered again at the central reference atom as well as the neighboring atoms N_{env} within a certain cutoff radius. The resulting density expression is

$$\rho_{SOAP}(\mathbf{R}) = \sum_{i=1}^{N_{env}} \exp(-\alpha |\mathbf{R} - \mathbf{R_i}|^2). \qquad (2.8)$$

The above is indeed closely related to the ACSF functions in Equation (2.1) and can be considered a three-dimensional generalization of the radial ACSF. A direct consequence of this observation is that while an overlap of two SOAP densities, ρ_{SOAP} and ρ'_{SOAP}, computed for two different central atoms in different structural environments, can be used to calculate a similarity between the atoms' local environments, this similarity would not be rotationally invariant. To overcome this issue, one of the SOAP densities needs to be integrated over all possible rotations, denoted as $\hat{R}$. Thus, the resulting rotationally invariant SOAP kernel is given by

$$k(\rho_{SOAP}, \rho'_{SOAP}) = \int d\hat{R} \left| \int \rho_{SOAP}(\mathbf{r}) \rho'_{SOAP}(\hat{\mathbf{R}}\mathbf{r}) d(\mathbf{r}) \right|^{n_{SOAP}}. \qquad (2.9)$$

The exponent n_{SOAP} is a parameter which is usually taken as 2. This parameter needs to be carefully chosen. For instance, $n_{SOAP} = 1$ is not a good choice since with this specific choice the order of the two integrals in the above expression can be interchanged as a result of which the rotational information is lost. Further, to enforce the self-similarity, the expression in Equation (2.9) needs to be normalized with a factor $1/\sqrt{k(\rho_{SOAP}, \rho_{SOAP}) \cdot k(\rho'_{SOAP}, \rho'_{SOAP})}$. The normalization ensures that within the adopted scheme the overlap of any given environment with itself is one. Representations based on the SOAP approach have been used extensively to successfully develop atomistic potentials for materials spanning a range of complexity, from bulk silicon to water clusters. While recent efforts have largely focused on configurational complexity for elemental or binary compounds, the efficacy of this approach on systems containing more than a few elements remains to be explored. This bottleneck, which is not specific to the SOAP representations and is shared by most machine learning based potentials, is one of the most exciting representations with potentially far reaching consequences that remains to be addressed.

Graph-Based Representations

More recently, graph-based representations have been successfully applied to represent molecules and solids [28, 73–75]. Faber et al. [74] benchmarked a range of different models on the QM9 data set [76] to show that the graph-based deep learning models generally out perform classical machine learning models. Chen et al. [75] developed physically intuitive MatErials Graph Network (MEGNet) models to demonstrate that the machine-learned element embeddings in these models encode periodic chemical trends and can be transfer-learned, i.e., a property model trained on a larger data set and given property can be used to improve predictions on other property models with smaller amounts of data. Xie and Grossman presented a crystal graph convolutional neural networks framework to learn material properties from the connection of atoms in the crystal. The method was shown to accurately predict a wide range of DFT-calculated properties (such as, formation energy, bandgap, Fermi energy, and different elastic moduli) of crystals with various structure types and compositions after being trained with $\sim 10^4$ data points. The ability to extract atomic contributions from local chemical environments to global properties and transfer learned knowledge from one set of properties or chemistries to the other make these representations a truly powerful tool for materials high throughput screening and design.

2.2.2   MESOSCOPIC REPRESENTATIONS

In the resolution-based hierarchy of materials representations, the mesoscopic descriptors represent the next rung of higher granularity. This class of representations captures a course-grained level of configurational details, while encoding relatively finer details than those captured by the bulk composition and elemental-constituent based macroscopic descriptors alone. More specifically, materials are described in terms of the basic building blocks (that is, clusters of atoms, fragments, or motifs) that eventually form the material. While the relative arrangement of atoms within each motif remains more-or-less uniform throughout the target materials class, the relative arrangements of different motifs with respect to each other largely dictates the observed variation in the target property going from one material to the other. In this regard, the descriptors at this level can be compared with the classic *cluster expansion* effective Hamiltonian approach in solid state physics [31, 32] or quantitative structure activity/property relationships (QSAR/QSPR) in cheminformatics [77].

With any classic QSAR/QSPR research, one describes the target class of compounds with a descriptor formed frequently by a combination of structure-agnostic empirical molecular properties (such as partition coefficients, molecular refraction, ionization potential, and molecular polarizability) and topological and coarse-grained structural features [78–80]. The selection of specific components of the descriptor is based on (i) its correlation with the property (activity) value of the target property and, if possible, (ii) its ability to provide some physical insight to the molecular/material behavior. In the quest of being able to predict outcomes of more expensive or time-consuming tests from such expensive and fast measurements/computations, the QSAR/QSPR approach has been tremendously successful. While the relationship between the number of atoms and boiling points observed in the series of *n*-alkanes represents a classic example of a QSPR model in organic chemistry [81], correlations on the set of *n*-alkanes and some other simple groups of compounds helped in building further more complicated *group-contribution* approaches to the problem [82]. In this regard, the popular approach of Van Krevelen and co-workers [30] established that a number of polymer properties such as glass transition temperatures, solubility parameters, and bulk moduli can be *explained* with the presence of chemical groups and combinations of different groups that form the polymer chain. In particular, the developed "group contribution" method was able to express various target properties as a linear weighted sum of the contribution coming from a set of diverse chemical groups such as $-CH_2-$, $-C_6H_4-$, $-COO-$, etc., that constituted the polymer repeat unit.

For several years, attention in this field has been focused on the topological descriptors which are molecular descriptors derived from information on connectivity and composition of a molecule. The creation, testing, and implementation of new mesoscale topological descriptors in organic chemistry has also been benefited from applications of graph theory and discrete mathematics [83, 84]. While the classical approaches based on mesoscopic materials representations are simple, chemically intuitive, and extensively used in the past to map chemo-structural features with a range of materials properties to come-up with practically useful correlation-based predictive models, most of the models assume that the target property is linearly related to the motif-parameters. This assumption can be particularly limiting when faced with highly nonlinear structure-property relationships in materials. Modern data-science methods built on mesoscopic descriptors allow us to address this limitation. In fact, a number of recent studies have demonstrated significant improvements in this direction.

As a part of a recent study [60, 85], which was geared toward accelerated discovery of novel polymer dielectrics, hundreds of polymer chemistries—all built from combination of seven basic organic polymer building blocks, such as $-CH_2$, -CO-, -CS-, -O-, -NH-, $-C_6H_4$-, and C_4H_2S, were considered. For these polymers, a selected set of properties relevant for dielectric applications, namely, the dielectric constant and bandgap, were computed using DFT computations [35, 60] inclusive of van der Waals interactions [86]. The generated DFT data was subsequently used to build efficient surrogate property prediction models of the target properties. The polymers were represented by a mesoscopic representation which accounted for the occurrence of a fixed set of molecular motifs in the polymer backbone in terms of their number fractions as well as their relative placements in the chain. It was demonstrated that the fingerprint based entirely on corresponding number fractions of the constituent groups and relative arrangement of pairs and triplets of the molecular motifs provides a uniform and seamless pathway to represent all polymers within this class, and the procedure can be easily generalized by considering higher-order fragments (that is, quadruples, quintuples, etc., of atom types) or by including new motifs. Finally, predictive mappings between the polymers and the properties were established using the KRR learning algorithm. These mappings were subsequently used to explore the Pareto optimal front (Chapter 5) of the bandgap-dielectric constant plots within the targeted class of polymers and led to the identification of promising polymers which were eventually synthesized and tested to be superior polymer dielectrics. The predictive tools developed as a part of this research have been made available online as a community resource and are constantly being updated to incorporate further developments [60].

The scope of molecular or material fragment-based mesoscopic representation is not restricted to quasi one-dimensional polymers and can readily be generalized to two- or three-dimensional materials. For instance, a similar materials representation strategy was employed to explore a vast chemical and property space presented by multi-layer superlattices of wurtzite AB-type crystalline solids stacked along the [0001] crystallographic direction [87]. The building blocks of these superlattices were drawn from a pool of 32 possible candidate binary crystalline

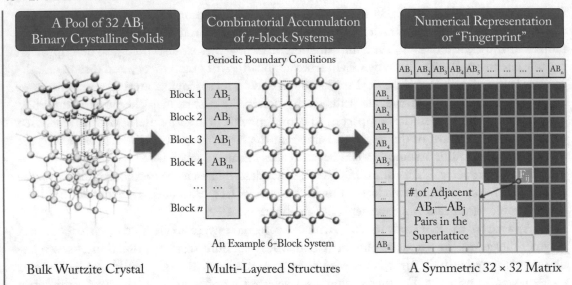

Figure 2.4: A schematic representation of vast chemical space available within multi-layer hetero-structures formed by compounds with a wurtzite crystal structures. Staring from a set 32 AB_i bulk compounds, multi-layer superlattices are formed in a combinatorial fashion. A simple mesoscopic motif-based numerical representation can be constructed by simply counting different possible AB_i–AB_j pairs appearing within a given superlattice, which can be used to learn/predict a wide range of properties for this class of compounds [87].

bulk solids, such as BN, BeO, SiC, AlN, GaN, ZnO, etc., that are known to occur either in the wurtzite crystal structure or in the zincblende ground state with a closely competing wurtzite phase. The chemical space made available by the superlattices of the AB_i compounds in this class is truly enormous, and the adopted scheme readily gives rise to a combinatorial explosion in terms of different compositional and configurational possibilities. For instance, if we just consider a *6-block* repeat unit, where each block can be populated by any of the 32 possibilities, a naive enumeration scheme leads to 6^{32} possibilities. Further accounting for double-counting owing to the transitional symmetry within the imposed periodic boundary conditions, although drastically reduces the unique possibilities, still leads to several million compounds. As a next step, to consider only physically meaningful and experimentally synthesizable superlattices, only those superlattices where none of the two bulk materials building the heterostructure have an average in-plane mismatch of more than 3% were considered. Allowing for superlattices formed by up to six independent AB_i chemistries lead to more than 1200 systems, for which a database of diverse properties such as lattice parameter, formation and interfacial energies, electronic bandgap, and elastic constants were computed with DFT computations. This work demonstrated that machine learning surrogate models, trained on carefully generated materials data, can provide an efficient alternative route toward materials property predictions and can be useful in identifying promising cases while searching for candidate materials with a set of pre-specified property requirements.

Mesoscopic representations have also been successfully developed and employed to identify novel compositions and the most probable crystal structures of ternary oxides (was $A_x B_y O_z$-type compounds) using a probabilistic model built on an experimental crystal structure database [36]. Such an approach required a descriptor that includes some aspects of the crystal structure information in combination with the composition information, that is, the elements that constitute the compound.

The three examples discussed above represent three classes of quasi-1D, quasi-2D, and 3D materials, respectively, where mesoscopic representations have been used to numerically represent the target material classes for a subsequent machine learning-based analysis. The utility of such fragment-based mesoscopic materials representations lies in the fact that any computable or experimentally measurable target property over a set of materials that vary in terms of both configuration and composition can be efficiently learned and predicted.

2.2.3   MACROSCOPIC REPRESENTATIONS

Macroscopic representations—constructed using easily accessible attributes of bulk composition and selected details of elemental species forming the compound—can be enormously useful in constructing low-dimensional decision rules or design plots concerning complex materials phenomena. One can find numerous examples where simple one- or two-dimensional macroscopic representations, based solely on chemical intuition and empirical knowledge, have had remarkable impact on the field (although these discoveries were not discussed in the context of materials informatics). Classical examples include Mendeleev's Periodic Table, Hume-Rothery's rules for miscibility of alloying elements in a solid solution [88], the Hall-Petch relationship connecting crystalline material's strength to the average grain size [89, 90], and a number of structure plots that use simple geometric features to describe formability of a set of chemistries across different crystal structures [91, 92]. While a laborious and careful manual analysis of experimental observations has played a key role in developing such design rules in the past, as we hinted in Chapter 1, modern data mining and statistical learning approaches provide an even more efficient and effective pathway toward automated knowledge discovery and rule induction from curated materials databases, especially for the situations where relationships are highly nonlinear and reside in a multi-dimensional feature space.

While high-level macroscopic descriptors are the simplest, often highly effective, easily accessible, and manually identified for certain situations where sufficient domain knowledge is available (as has been the case traditionally), for a general problem where the choice of the relevant descriptors is not obvious, an unbiased and automated coarse level descriptors selection is often required. Further, if the multi-variate relationships are anticipated to be highly nonlinear, the starting "linear" or primary coarse-level descriptors can be combined with a set of preselected nonlinear algebraic functions to enlarge the starting pool of possibilities. Starting from a moderate number of primary descriptors and a limited set of algebraic functions and only allowing physically meaningful combinations of derived descriptors (for instance, not considering additions and subtractions of descriptors with different units), one can quickly accumulate rather large sets of descriptors, containing several millions or even billions candidates. This large nonlinear feature space can subsequently be "mined" for the descriptors which are highly correlated with the targeted property [10, 93, 94]. Finding this hidden structure in the vast feature sets requires dedicated methodology based on approaches from statistical learning, information science, and applied mathematics.

One such approach, known as the least absolute shrinkage and selection operator (LASSO), was recently demonstrated to be highly effective for finding physically meaningful low-dimensional descriptors for materials classification and regression problems [7, 10]. Within the LASSO framework [95], we start with a training set of n data $\mathcal{D} = \{(x_i, y_i), |i = 1, \ldots, n\}$ where an x_i is a vector of the values of the D macroscopic features and the y_i are the labels/target properties for material i. Usually, the dimensionality D of the feature space is large, and the goal is to find an interpretable explicit analytical functional form $y' = f(x) = x \cdot c$, where c is a column vector of scalar coefficients that needs to be determined. It is required that the predicted properties y' depend only on a small number of input parameters; that is, most of the components in c vanish, providing a sparse solution where only a subset of the initial feature set x survives. The most straightforward approach is to minimize a regularized least-squared expression,

$$\underset{c}{\mathrm{argmin}} \| y - x \cdot c \|_2^2 + \lambda \mathcal{L}(c),$$

where $\| \cdot \|_2^2$ is the squared Euclidian norm and $\mathcal{L}(c)$ is a regularization term, with λ being a hyper-parameter controlling its strength. Note that setting the regularization term to zero or $\|c\|_2^2 = \sum_j c_j^2$, simply leads to the well-known linear least-squared fit and ridge regression models. Since we are specifically looking for low-dimensional or *compact* patterns in the data $\mathcal{D}$, where the target property should only depend on a selected few components of x, ideally the best-performing sparse solutions of the above equations are desired. Therefore, in principle, one needs to minimize an l_0-norm along with the least square expression above, that is, $\mathcal{L}(c) = \|c\|_0 = \#(j : c_j \neq 0)$ with a $\#$ representing the number of elements in the set. However, this is an NP-hard problem (that is, non-deterministic polynomial-time hard) which quickly becomes computationally intractable as the dimension D grows. A practical solution is to resort to an l_1-norm with $\mathcal{L}(c) = \|c\|_1 = \sum_j |c_j|$ that unlike the l_0-norm regularization leads to a convex problem and the geometry of the solution still prefers a sparse solution. Furthermore, by tuning the strength of the regularization via λ, it is possible to control the sparsity of the solution.

In a pioneering work, a modification of the LASSO approach was used by Ghiringhelli and co-workers [7] to classify binary octet insulators into tendencies for the formation of rock salt vs. zinc blende structures to demonstrate that the mentioned compressed sensing framework can indeed find novel and physically meaningful patterns in materials data. However, in moving beyond the showcased example problem of predicting the ground-state crystal structure of octet binaries semiconductors, it turned out that the method can quickly become highly inefficient when faced with truly large feature spaces, particularly in situations where little understanding of the underlying physical mechanisms is available. The method also tends to break down (that is, leads to sub-optimal solutions) when the features are correlated (which is not unlikely particularly when one starts out with a truly vast feature set).

To address the shortcomings of the LASSO-based approach, a more sophisticated and robust approach, namely, the sure independent screening and sparsifying operator (SISSO) method, was proposed [93, 94] which efficiently deals with feature sets containing billions or more features and does not suffer when features are correlated. These recent developments offer new opportunity to inspect and discover previously unknown analytical relationships in form of design rules or decision maps and suggest means to test the generalization ability of the identified design rules. Some prominent examples of the mentioned compressed sensing descriptor design include novel classification maps for crystal structures of solids [7, 96, 97], predictions of the bandgap in solids [12], low-dimensional representation for alloy formation energy [98], and construction of simpler phenomenological models for complex materials phenomena [10, 11, 99].

These advances enabled the community to take a fresh look at available materials databases for previously unknown chemical insights and hidden trends in the data. Along these lines, another exciting example of crystal structure classification, in addition to the rock salt vs. zinc blende classification of octet-binary semiconductors, was recently proposed by Bialon and co-workers [13]. This study considered a set of 2105 known experimental $A_x B_y$ type compounds, containing sp-block elements at the A sub-lattice and transition metals forming the B sub-lattice, with an aim to learn a structural map that would classify the compounds in all 64 different possible prototypical crystal structures. After searching a set of 1.7×10^5 nonlinear descriptors formed by physically meaningful functions of primary coarse-level descriptors such as band-filling, atomic volume, and different electronegativity scales of the sp and d elements, the authors identified a set of 3 optimal descriptors that classified all the experimentally known training examples available from the Pearson's Crystal Database [100] with an 86% probability of predicting the correct crystal structure. Further, the model was demonstrated to be robust with respect to different subsets of training data and exhibited a confidence of better than 98% that the correct crystal structure is among three predicted crystal structures.

The role of identified macroscopic representations for descriptors has not been limited to explaining the underlying chemical trends in materials data. In fact, several examples in the recent literature have demonstrated predictive capabilities of the statistical approach. In many cases, materials informatics guided efforts have led to identification of novel compounds that have been either synthesized/tested or verified by accurate and expensive first-principles based quantum mechanical simulations to validate the machine learning predictions. For instance, Oliynyk et al. [101] recently employed a set of elemental descriptors to train a machine-learning model, built on a random forest algorithm [102] (Chapter 3), with an aim to accelerate the search for Heusler compounds. After the model was trained on available crystallographic data from Pearson's Crystal Database [100] and the ASM Alloy Phase Diagram Database [103], the model was used to evaluate the probabilities at which ternary compounds with the formula AB_2C will adopt the Heusler structures. This approach was exceptionally successful in distinguishing between Heusler and non-Heusler compounds with an impressive true positive rate of 94%, including the prediction of unknown compounds and flagging erroneously assigned entries in the literature and in crystallographic databases. As a proof of concept, a set of 12 novel predicted Gallides compounds, with formulae MRu_2Ga and RuM_2Ga, where M = Ti, V, Cr, Mn, Fe and Co, were synthesized and confirmed to be Heusler compounds.

In a different example, Kim et al. [10] used LASSO-based feature selection to discover a general and analytical phenomenological model for the intrinsic dielectric breakdown field of insulators. This is a property that represents the theoretical limit of applied electric field at which an insulator material *breaks-down* and starts behaving as a conductor. The intrinsic dielectric breakdown strength of a compound, determined purely by the chemistry of the material (that is, the elements the material is composed of, the atomic-level structure, and the bonding), can be enormously time and computation resource intensive to access from first-principles based quantum mechanical simulations, and therefore cheaper surrogate models for the property are highly desirable and practically useful. With this motivation, starting from a benchmark dataset of 83 sp-bonded binary octet dielectric materials, containing alkali metal halides, transition metal halides, alkaline earth metal chalcogenides, transition metal oxides, and group III, II-VI, I-VII semiconductors, Kim et al. developed trained and validated informatics models that revealed key correlations and analytical relationships between the breakdown field and other easily accessible material properties such as the bandgap and the phonon cutoff frequency (Figure 2.5). The resulting phenomenological model was shown to be general and was later employed to systematically screen perovskite compounds with high breakdown strength. This study identified that boron-containing compounds are of particular interest, some of which were predicted to exhibit remarkable intrinsic breakdown strength of 1 GV/m. These predictions were subsequently confirmed using first principles computations [11].

Another promising and potentially impactful avenue for machine learning models built on macroscopic descriptors is to learn the tendency for metallic vs. insulating behavior of a given chemistry (that is, given its composition and crystal structure). The fact that the well-established standard first-principles approach or DFT has a well-known deficiency in predicting this tendency [104, 105], making the statistical route for the bandgap learning even more crucial and practically useful. Furthermore, the materials bandgap represents a key property central to diverse technological application including electronic, energy harvesting, energy storage, and catalysis and often serves as a crucial screening parameter in rational design of functional materials [6, 106–115]. Therefore, it is not surprising that numerous efforts have already been made to identify suitable macroscopic descriptors that can qualitatively or quantitatively describe the materials bandgap. To be most effective, all of these studies have targeted a pre-defined and limited

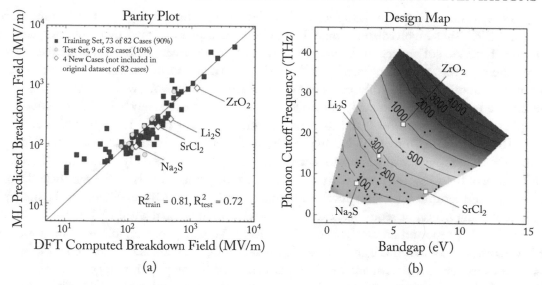

Figure 2.5: (a) Parity plot comparing DFT-computed intrinsic dielectric breakdown field against predicted intrinsic breakdown field for a set of 82 binary octet AB-type compounds using randomly selected 90% training and 10% test sets, as well as for 4 new compounds not included during the model development stage. (b) A design map for the prediction of intrinsic dielectric breakdown field using the two most relevant features identified in the machine learning based surrogate model development, namely the bandgap and the phonon cutoff frequency (that is, the highest possible phonon frequency). The contour lines and colors on the 2D map represent the intrinsic dielectric breakdown field strength in MV/m. Adapted from [10] with permission; Copyright (2017) American Chemical Society.

class of materials such as binary $A_x B_y$ type compounds, chalcopyrite compounds, MAXenes, or complex perovskites [12, 16, 116, 117].

Within a given class of materials, an machine learning model that can predict the bandgap, given just the composition, can be of enormous practical value and can be readily employed as a filter in a first line of screening to quickly narrow down the set of promising compounds that need to be explored in further detail.

Rajan and coworkers [116] use experimentally available bandgaps of ABC_2 chalcopyrite compounds to train regression models with electronegativity, atomic number, melting point, pseudopotential radii, valence for each of the A, B, and C elements as features. Just with the gross-level elemental features, the developed machine learning models predicted the experimental bandgaps with a moderate accuracy. In a different study, Pilania et al. [12] used a database consisting of accurate bandgaps of 1300 $AA'BB'O_6$ type double perovskites to train a kernel ridge regression machine learning model for efficient predictions of the bandgaps. A set of descriptors with increasing complexity were identified via searching a large part of feature space using LASSO that involved more than 1.2 million descriptors formed by combining simple algebraic functions of elemental features such as electronegativities, ionization potentials, electronic energy levels, and valence orbital radii of the constituent atomic species. One of the most important chemical insights that came out of the learning framework was that the bandgap in the double perovskites is primarily controlled (and therefore effectively learned) by the lowest occupied energy levels of the A-site elements and electronegativities of the B-site elements.

The bandgap learning models also provide an efficient and effective pathway toward learning bandgaps computed using more advanced and highly computationally demanding "beyond-DFT" level of theories. The predictive power of bandgap leaning models can be significantly improved by augmenting the coarse-level elemental descriptors with Kohn-Sham (KS) bandgaps computed using a local or semi-local exchange correlation functional within DFT. For instance, Lee et al. [117] reported an excellent quasi-particle GW bandgaps prediction model for a set of 270 inorganic compounds using KS bandgaps as features in the machine learning. However, since for every new prediction one needs to perform a DFT computation to get a KS bandgap, this approach can be computationally demanding for large chemical space explorations. An alternative multi-fidelity learning approach (Chapter 6) was recently proposed to overcome this deficiency, where one simultaneously trains two learning models using just the elemental features to learn a low-fidelity bandgap model (for example, KS bandgaps computed with conventional DFT) and a difference model for the high and low fidelities (for example, the difference between the GW and the KS bandgaps) [118]. Once trained, the multi-fidelity machine learning does not necessarily require bandgaps computed at a lower (that is, less accurate but computationally cheaper) level of theory data to make predictions of bandgaps from a higher level of theory. The subject of multi-fidelity information-fusion is covered in further detail in Chapter 6.

While regression models for bandgap prediction, such as those discussed above, can provide quantitative information for insulating materials, a capability to efficiently classify a given set of materials as metals or insulators is equally desirable. In fact, in many high-throughput studies targeted toward finding materials for a specific application metal vs. insulator classification often constitutes a step in the over all screening process. Recently, Ouyang et al. applied a SISSO-based approach to the showcase example of predicting the metal-insulator classification of binary compounds using experimental data on the bandgaps [93]. Accurate, transparent, and predictive models in terms of low dimensional analytical descriptors were found which led to almost perfect classification (with 99.0% accuracy) of metal/nonmetal for 299 materials (Figure 2.6a). Further, for the metal-insulator classification model, the predictive capabilities are tested beyond the training data by using the model to rediscover the available pressure-induced insulator to metal transitions. It was shown that, in addition to finding known pressure-induced metal to insulator transitions, the model allows for the prediction of yet unknown transition candidates, ripe for experimental validation (Figure 2.6b). General (that is, applicable to any

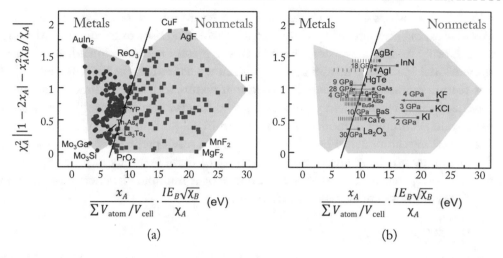

Figure 2.6: (a) A metal-nonmetal classification map for 299 for a set of 299 materials enabled by a 2D descriptor identified by SISSO. Symbols used in descriptors: χ, Pauling electronegativity; IE, ionization energy; x, atomic composition; $\sum V_{atom}/V_{cell}$, packing fraction. Red circles, blue squares, and open blue squares represent metals, nonmetals, and the three erroneously characterized nonmetals, respectively. (b) Reproduction of pressure-induced insulator-metal transitions (red arrows), of materials that remain insulators upon compression (blue arrows), and computational predictions at step of 1 GPa (green bars). Reproduced from [93], with permission. Copyright (2018) American Physical Society.

materials property) and specifically designed for relatively small dataset problems, such models built on macroscopic features represent an effective tool for automatic materials development.

The enterprise of developing materials representations for either optimization of existing materials or to design novel functional materials remains an interesting, challenging, and active area in materials informatics. In this chapter, we have covered a broader spectrum of features's landscape possible in materials informatics problems. A hierarchical classification of features was presented to emphasize that different application-specific predictive models may require features at varying levels of granularity. A number of examples from the recent literature were presented as practical illustrations. Certain finer aspects of feature selection and feature extraction were not covered here. In particular, dimensionality reduction methods are not explicitly covered. It is instructive to understand as to why this step is necessary. Perhaps, the two most important reasons are to reduce the curse of dimensionality and to improve computational efficiency. Although the curse of dimensionality poses a serious issue for machine learning applications in general, in materials science it is possible to circumvent this problem by realizing the fact that in real situations the available data is often confined to a region of the feature space with a significantly lower effective dimensionality. The role of various dimensionality reduction techniques is to identify these selective dimensions which carry most of the relevant information. Typically, this is done by simply identifying and discarding those variables which do not contribute toward a targeted learning task or by finding a transformation from the present n-dimensional feature space to a lower dimensional space. The former approach is known as feature selection, while the latter is known as feature extraction. It should be noted that feature selection or dimensionality reduction is not necessarily required as a preprocessing step for classification or regression algorithms to perform well. A number of modern machine learning algorithms use built-in regularization or feature selection techniques to handle overfitting, as a result preprocessing to remove unnecessary features may not be required. Some examples of these methods include support vector machines, LASSO, tree-based classifiers, or average over multiple classifiers (ensemble methods). The last two topics are discussed in Chapter 3.

2.3   REFERENCES

[1] C. M. Bishop. *Pattern Recognition and Machine Learning*. Springer-Verlag, New York, 2006. 25

[2] John F. Dobson, Beyond pairwise additivity in London dispersion interactions. *International Journal of Quantum Chemistry*, 114.18 (20:14):1157–1161. 34

[3] T. Mueller, A. G. Kusne, and R. Ramprasad. Machine learning in materials science: Recent progress and emerging applications. *Reviews in Computational Chemistry*, 29:186, 2016. DOI: 10.1002/9781119148739.ch4 25

[4] T. Lookman, F. J. Alexander, and K. Rajan, Eds. *Information Science for Materials Discovery and Design*, volume 225. Springer Series in Materials Science, Heidelberg, 2016. DOI: 10.1007/978-3-319-23871-5

[5] R. Ramprasad, R. Batra, G. Pilania, A. Mannodi-Kanakkithodi, and C. Kim. Machine learning in materials informatics: Recent applications and prospects. *NPJ Computational Materials*, 3(1):54, 2017. DOI: 10.1038/s41524-017-0056-5 25

[6] G. Pilania, K. J. McClellan, C. R. Stanek, and B. P. Uberuaga. Physics-informed machine learning for inorganic scintillator discovery. *Journal of Chemical Physics*, 148(24):241729, 2018. DOI: 10.1063/1.5025819 25, 44

[7] L. M. Ghiringhelli, J. Vybiral, S. V. Levchenko, C. Draxl, and M. Scheffler. Big data of materials science: Critical role of the descriptor. *Physical Review Letters*, 114(10):105503, 2015. DOI: 10.1103/physrevlett.114.105503 25, 42, 43

[8] A. Seko, A. Togo, H. Hayashi, K. Tsuda, L. Chaput, and I. Tanaka. Prediction of low-thermal-conductivity compounds with first-principles anharmonic lattice-dynamics calculations and Bayesian optimization. *Physical Review Letters*, 115(20):205901, 2015. DOI: 10.1103/physrevlett.115.205901 26

[9] D. Xue, P. V. Balachandran, J. Hogden, J. Theiler, D. Xue, and T. Lookman. Accelerated search for materials with targeted properties by adaptive design. *Nature Communications*, 7:11241, 2016. DOI: 10.1038/ncomms11241

[10] C. Kim, G. Pilania, and R. Ramprasad. From organized high-throughput data to phenomenological theory using machine learning: The example of dielectric breakdown. *Chemistry Materials*, 28(5):1304, 2016. DOI: 10.1021/acs.chemmater.5b04109 42, 43, 44, 45

[11] C. Kim, G. Pilania, and R. Ramprasad. Machine learning assisted predictions of intrinsic dielectric breakdown strength of ABX_3 perovskites. *Journal of Physical Chemistry C*, 120(27):14575, 2016. DOI: 10.1021/acs.jpcc.6b05068 43, 44

[12] G. Pilania, A. Mannodi-Kanakkithodi, B. P. Uberuaga, R. Ramprasad, J. E. Gubernatis, and T. Lookman. Machine learning bandgaps of double perovskites. *Scientific Reports*, 6:19375, 2016. DOI: 10.1038/srep19375 43, 45

[13] A. F. Bialon, T. Hammerschmidt, and R. Drautz. Three-parameter crystal-structure prediction for sp-d-valent compounds. *Chemistry Materials*, 28(8):2550, 2016. DOI: 10.1021/acs.chemmater.5b04299 43

[14] G. Pilania, P. V. Balachandran, J. E. Gubernatis, and T. Lookman. Predicting the formability of ABO_3 perovskite solids via machine learning study. *Acta Crystallographica B*, 71:507, 2017. DOI: 10.1107/S2052520615013979

[15] Z. Li, X. Ma, and H. Xin. Feature engineering of machine-learning chemisorption models for catalyst design. *Catalysis Today*, 280:232, 2017. DOI: 10.1016/j.cattod.2016.04.013

[16] A. C. Rajan, A. Mishra, S. Satsangi, R. Vaish, H. Mizuseki, K.-R. Lee, and A. K. Singh. Machine-learning-assisted accurate band gap predictions of functionalized MXene. *Chemistry Materials*, 30(12):4031, 2018. DOI: 10.1021/acs.chemmater.8b00686 26, 45

[17] M. Rupp, A. Tkatchenko, K.-R. Müller, and O. A. von Lilienfeld. Fast and accurate modeling of molecular atomization energies with machine learning. *Physical Review Letters*, 108(5):058301, 2012. DOI: 10.1103/physrevlett.108.058301 26, 30, 33

[18] M. Rupp. Machine learning for quantum mechanics in a nutshell. *International Journal of Quantum Chemistry*, 115(16), 1058–1073, 2015. 35

[19] K. Hansen, G. Montavon, F. Biegler, S. Fazli, M. Rupp, M. Scheffler, O. A. von Lilienfeld, A. Tkatchenko, and K.-R. Müller. Assessment and validation of machine learning methods for predicting molecular atomization energies. *Journal of Chemical Theory and Computation*, 9(8):3404, 2013. DOI: 10.1021/ct400195d 26, 34

[20] F. Faber, A. Lindmaa, O. A. von Lilienfeld, and R. Armiento. Crystal structure representations for machine learning models of formation energies. *International Journal of Quantum Chemistry*, 115(16):1094, 2015. DOI: 10.1002/qua.24917 33

[21] J. Behler. Constructing high-dimensional neural network potentials: A tutorial review. *International Journal of of Quantum Chemistry*, 115(16):1032, 2015. DOI: 10.1002/qua.24890

[22] K. Hansen, F. Biegler, R. Ramakrishnan, W. Pronobis, O. A. von Lilienfeld, K.-R. Müller, and A. Tkatchenko. Machine learning predictions of molecular properties: Accurate many-body potentials and nonlocality in chemical space. *Journal of Physical Chemistry Letters*, 6(12):2326, 2015. DOI: 10.1021/acs.jpclett.5b00831 34, 35

[23] J. Behler. Perspective: Machine learning potentials for atomistic simulations. *Journal of Chemical Physics*, 145(17):170901, 2016. DOI: 10.1063/1.4966192 32

[24] T. D. Huan, R. Batra, J. Chapman, S. Krishnan, L. Chen, and R. Ramprasad. A universal strategy for the creation of machine learning-based atomistic force fields. *NPJ Computational Materials*, 3(1):37, 2017. DOI: 10.1038/s41524-017-0042-y 32, 33

[25] N. Artrith, A. Urban, and G. Ceder. Efficient and accurate machine-learning interpolation of atomic energies in compositions with many species. *Physical Review B*, 96(1):014112, 2017. DOI: 10.1103/physrevb.96.014112

[26] C. Chen, Z. Deng, R. Tran, H. Tang, I.-H. Chu, and S. P. Ong. Accurate force field for molybdenum by machine learning large materials data. *Physical Review Materials*, 1(4):043603, 2017. DOI: 10.1103/physrevmaterials.1.043603

[27] H. Zong, G. Pilania, X. Ding, G. J. Ackland, and T. Lookman. Developing an interatomic potential for martensitic phase transformations in zirconium by machine learning. *NPJ Computational Materials*, 4(1):48, 2018. DOI: 10.1038/s41524-018-0103-x

[28] T. Xie and J. C. Grossman. Crystal graph convolutional neural networks for an accurate and interpretable prediction of material properties. *Physical Review Letters*, 120(14):145301, 2018. DOI: 10.1103/physrevlett.120.145301 37

[29] A. Chandrasekaran, D. Kamal, R. Batra, C. Kim, L. Chen, and R. Ramprasad. Solving the electronic structure problem with machine learning. *NPJ Computational Materials*, 5(1):22, 2019. DOI: 10.1038/s41524-019-0162-7 26, 29, 33

[30] D. W. Van Krevelen and K. T. Nijenhuis. *Properties of Polymers: Their Correlation with Chemical Structure; Their Numerical Estimation and Prediction from Additive Group Contributions*. Elsevier, Berlin, 2009. DOI: 10.1016/C2009-0-05459-2 27, 38

[31] J. M. Sanchez, F. Ducastelle, and D. Gratias. Generalized cluster description of multicomponent systems. *Physica A*, 128(1-2):334, 1984. DOI: 10.1016/0378-4371(84)90096-7 27, 38

[32] D. J. De Fontaine. Cluster approach to order-disorder transitions in alloys. In H. Ehrenreich and D. Turnbull, Eds., *Solid State Physics*, volume 47, p. 33. Academic Press, Cambridge, MA, 1994. DOI: 10.1063/1.31877 27, 38

[33] C. Nantasenamat, C. Isarankura-Na-Ayudhya, T. Naenna, and V. Prachayasittikul. A practical overview of quantitative structure-activity relationship. *EXCLI*, 8:74, 2009. 27

[34] G. Pilania, C. Wang, X. Jiang, S. Rajasekaran, and R. Ramprasad. Accelerating materials property predictions using machine learning. *Scientific Reports*, 3:2810, 2013. DOI: 10.1038/srep02810 27

[35] T. D. Huan, A. Mannodi-Kanakkithodi, C. Kim, V. Sharma, G. Pilania, and R. Ramprasad. A polymer dataset for accelerated property prediction and design. *Scientific Data*, 3:160012, 2016. DOI: 10.1038/sdata.2016.12 39

[36] G. Hautier, C. C. Fischer, A. Jain, T. Mueller, and G. Ceder. Finding nature's missing ternary oxide compounds using machine learning and density functional theory. *Chemistry Materials*, 22(12):3762, 2010. DOI: 10.1021/cm100795d 41

[37] C. Kim, A. Chandrasekaran, T. D. Huan, D. Das, and R. Ramprasad. Polymer genome: A data-powered polymer informatics platform for property predictions. *Journal of Physical Chemistry C*, 122(31):17575, 2018. DOI: 10.1021/acs.jpcc.8b02913

[38] A. Mannodi-Kanakkithodi, A. Chandrasekaran, C. Kim, T. D. Huan, G. Pilania, V. Botu, and R. Ramprasad. Scoping the polymer genome: A roadmap for rational polymer dielectrics design and beyond. *Materials Today*, 21(7):785, 2018. DOI: 10.1016/j.mattod.2017.11.021 27, 30

[39] P. Hohenberg and W. Kohn. Inhomogeneous electron gas. *Physical Review*, 136(3B):B864, 1964. DOI: 10.1103/physrev.136.b864 28

[40] W. Kohn and L. J. Sham. Self-consistent equations including exchange and correlation effects. *Physical Review*, 140(4A):A1133, 1965. DOI: 10.1103/physrev.140.a1133 28

[41] J. C. Snyder, M. Rupp, K. Hansen, K.-R. Müller, and K. Burke. Finding density functionals with machine learning. *Physical Review Letters*, 108(25):253002, 2012. DOI: 10.1103/physrevlett.108.253002 29

[42] G. Montavon, M. Rupp, V. Gobre, A. Vazquez-Mayagoitia, K. Hansen, A. Tkatchenko, K.-R. Müller, and O. A. von Lilienfeld. Machine learning of molecular electronic properties in chemical compound space. *New Journal of Physics*, 15(9):095003, 2013. DOI: 10.1088/1367-2630/15/9/095003

[43] K. T. Schütt, H. Glawe, F. Brockherde, A. Sanna, K. R. Müller, and E. K. U. Gross. How to represent crystal structures for machine learning: Towards fast prediction of electronic properties. *Physical Review B*, 89(20):205118, 2014. DOI: 10.1103/physrevb.89.205118

[44] F. Brockherde, L. Vogt, L. Li, M. E. Tuckerman, K. Burke, and K.-R. Müller. Bypassing the Kohn-Sham equations with machine learning. *Nature Communications*, 8(1):872, 2017. DOI: 10.1038/s41467-017-00839-3

[45] K. Schütt, P.-J. Kindermans, H. E. S. Felix, S. Chmiela, A. Tkatchenko, and K.-R. Müller. SchNet: A continuous-filter convolutional neural network for modeling quantum interactions. In *Advances in Neural Information Processing Systems*, p. 991, 2017.

[46] K. T. Schütt, H. E. Sauceda, P.-J. Kindermans, A. Tkatchenko, and K.-R. Müller. SchNet—A deep learning architecture for molecules and materials. *Journal of Chemical Physics*, 148(24):241722, 2018. DOI: 10.1063/1.5019779 29

[47] H. Huo and M. Rupp. Unified representation of molecules and crystals for machine learning. https://arxiv.org/abs/1704.06439, 2017. 30, 36

[48] A. P. Bartók, M. C. Payne, R. Kondor, and G. Csányi. Gaussian approximation potentials: The accuracy of quantum mechanics, without the electrons. *Physical Review Letters*, 104(13):136403, 2010. DOI: 10.1103/physrevlett.104.136403 30, 36

[49] J. Behler and M. Parrinello. Generalized neural-network representation of high-dimensional potential-energy surfaces. *Physical Review Letters*, 98:146401, 2007. DOI: 10.1103/physrevlett.98.146401 30, 31, 33

[50] V. Botu and R. Ramprasad. Learning scheme to predict atomic forces and accelerate materials simulations. *Physical Review B*, 92(9):094306, 2015. DOI: 10.1103/physrevb.92.094306 30, 32, 33

[51] O. A. von Lilienfeld, R. Ramakrishnan, M. Rupp, and A. Knoll. Fourier series of atomic radial distribution functions: A molecular fingerprint for machine learning models of quantum chemical properties. *International Journal of Quantum Chemistry*, 115(16):1084, 2015. DOI: 10.1002/qua.24912

[52] A. P. Thompson, L. P. Swiler, C. R. Trott, S. M. Foiles, and G. J. Tucker. Spectral neighbor analysis method for automated generation of quantum-accurate interatomic potentials. *Journal of Computational Physics*, 285:316, 2015. DOI: 10.1016/j.jcp.2014.12.018 30, 36

[53] R. M. Martin. *Electronic Structure: Basic Theory and Practical Methods*. Cambridge University Press, Cambridge, 2004. DOI: 10.1017/cbo9780511805769 30

[54] D. C. Rapaport. *The Art of Molecular Dynamics Simulation*. Cambridge University Press, Cambridge, 2004. DOI: 10.1017/cbo9780511816581 30

[55] M. S. Daw and M. I. Baskes. Embedded-atom method: Derivation and application to impurities, surfaces, and other defects in metals. *Physical Review B*, 29(12):6443, 1984. DOI: 10.1103/physrevb.29.6443 30

[56] R. Drautz and D. G. Pettifor. Valence-dependent analytic bond-order potential for transition metals. *Physical Review B*, 74(17):174117, 2006. DOI: 10.1103/physrevb.74.174117

[57] T. Liang, T.-R. Shan, Y.-T. Cheng, B. D. Devine, M. Noordhoek, Y. Li, Z. Lu, S. R. Phillpot, and S. B. Sinnott. Classical atomistic simulations of surfaces and heterogeneous interfaces with the charge-optimized many body (COMB) potentials. *Materials Science Engineering R*, 74(9):255, 2013. DOI: 10.1016/j.mser.2013.07.001

[58] T. P. Senftle, S. Hong, M. M. Islam, S. B. Kylasa, Y. Zheng, Y. K. Shin, C. Junkermeier, R. Engel-Herbert, M. J. Janik, and H. M. Aktulga. The ReaxFF reactive force-field: Development, applications and future directions. *NPJ Computational Materials*, 2:15011, 2016. DOI: 10.1038/npjcompumats.2015.11 30

[59] R. Ramakrishnan and O. A. von Lilienfeld. Many molecular properties from one kernel in chemical space. *CHIMIA International Journal of Chemistry*, 69(4):182, 2015. DOI: 10.2533/chimia.2015.182 30

[60] A. Mannodi-Kanakkithodi, G. Pilania, T. D. Huan, T. Lookman, and R. Ramprasad. Machine learning strategy for accelerated design of polymer dielectrics. *Scientific Reports*, 6:20952, 2016. DOI: 10.1038/srep20952 30, 39

[61] J. Behler. Representing potential energy surfaces by high-dimensional neural network potentials. *Journal of Physics: Condensed Matter*, 26(18):183001, 2014. DOI: 10.1088/0953-8984/26/18/183001 32

[62] M. Hellström and J. Behler. Neural network potentials in materials modeling. In W. Andreoni and S. Yip, Eds., *Handbook of Materials Modeling: Methods: Theory and Modeling*, p. 1, Springer-Verlag, Heidelberg, 2018. DOI: 10.1007/978-3-319-42913-7_56-1

[63] C. M. Handley and J. Behler. Next generation interatomic potentials for condensed systems. *European Physical Journal of B*, 87(7):152, 2014. DOI: 10.1140/epjb/e2014-50070-0

[64] V. Quaranta, J. Behler, and M. Hellström. Structure and dynamics of the liquid—water/zinc-oxide interface from machine learning potential simulations. *Journal of Physical Chemistry C*, 123(2):1293, 2018. DOI: 10.1021/acs.jpcc.8b10781

[65] C. Schran, F. Uhl, J. Behler, and D. Marx. High-dimensional neural network potentials for solvation: The case of protonated water clusters in helium. *Journal of Chemical Physics*, 148(10):102310, 2018. DOI: 10.1063/1.4996819 32

[66] J. Behler. Atom-centered symmetry functions for constructing high-dimensional neural network potentials. *Journal of Chemical Physics*, 134(7):074106, 2011. DOI: 10.1063/1.3553717 32

[67] V. Botu, R. Batra, J. Chapman, and R. Ramprasad. Machine learning force fields: Construction, validation, and outlook. *Journal of Physical Chemistry C*, 121(1):511, 2016. DOI: 10.1021/acs.jpcc.6b10908 33

[68] J. J. P. Stewart. Optimization of parameters for semiempirical methods V: Modification of NDDO approximations and application to 70 elements. *Journal of Molecular Modeling*, 13(12):1173, 2007. DOI: 10.1007/s00894-007-0233-4 33

[69] S. W. Benson. III-Bond energies. *Journal of Chemical Education*, 42(9):502, 1965. DOI: 10.1021/ed042p502 33

[70] B. Huang and O. A. von Lilienfeld. Communication: Understanding molecular representations in machine learning: The role of uniqueness and target similarity. *Journal of Chemical Physics*, 145(16):161102, 2016. DOI: 10.1063/1.4964627 33

[71] S. A. Dianat and R. M. Rao. Fast algorithms for phase and magnitude reconstruction from bispectra. *Optical Engineering*, 29(5):504, 1990. DOI: 10.1117/12.55619 36

[72] A. P. Bartók, R. Kondor, and G. Csányi. On representing chemical environments. *Physical Review B*, 87(18):184115, 2013. DOI: 10.1103/physrevb.87.184115 36

[73] S. Kearnes, K. McCloskey, M. Berndl, V. Pande, and P. Riley. Molecular graph convolutions: Moving beyond fingerprints. *Journal of Computer–Aided Molecular Design*, 30(8):595, 2016. DOI: 10.1007/s10822-016-9938-8 37

[74] F. A. Faber, L. Hutchison, B. Huang, J. Gilmer, S. S. Schoenholz, G. E. Dahl, O. Vinyals, S. Kearnes, P. F. Riley, and O. A. von Lilienfeld. Prediction errors of molecular machine learning models lower than hybrid DFT error. *Journal of Chemical Theory and Computation*, 13(11):5255, 2017. DOI: 10.1021/acs.jctc.7b00577 37

[75] C. Chen, W. Ye, Y. Zuo, C. Zheng, and S. P. Ong. Graph networks as a universal machine learning framework for molecules and crystals. *arXiv:1812.05055*, 2018. DOI: 10.1021/acs.chemmater.9b01294 37

[76] R. Ramakrishnan, P. O. Dral, M. Rupp, and O. A. von Lilienfeld. Quantum chemistry structures and properties of 134 kilo molecules. *Scientific Data*, 1:140022, 2014. DOI: 10.1038/sdata.2014.22 37

[77] W. M. Berhanu, G. G. Pillai, A. A. Oliferenko, and A. R. Katritzky. Quantitative structure—activity/property relationships: The ubiquitous links between cause and effect. *ChemPlusChem*, 77(7):507, 2012. DOI: 10.1002/cplu.201200038 38

[78] J. Dong, D.-S. Cao, H.-Y. Miao, S. Liu, B.-C. Deng, Y.-H. Yun, N.-N. Wang, A.-P. Lu, W.-B. Zeng, and A. F. Chen. ChemDes: An integrated web-based platform for molecular descriptor and fingerprint computation. *Journal of Cheminformatics*, 7(1):60, 2015. DOI: 10.1186/s13321-015-0109-z 38

[79] H. Moriwaki, Y.-S. Tian, N. Kawashita, and T. Takagi. Mordred: A molecular descriptor calculator. *Journal of Cheminformatics*, 10(1):4, 2018. DOI: 10.1186/s13321-018-0258-y

[80] J. C. Stålring, L. A. Carlsson, P. Almeida, and S. Boyer. AZOrange-High performance open source machine learning for QSAR modeling in a graphical programming environment. *Journal of Cheminformatics*, 3(1):28, 2011. DOI: 10.1186/1758-2946-3-28 38

[81] F. C. Whitmore. *Organic Chemistry*. Dover Publications, New York, 1951. 38

[82] D. E. Needham, I. C. Wei, and P. G. Seybold. Molecular modeling of the physical properties of alkanes. *Journal of the American Chemical Society*, 110(13):4186, 1988. DOI: 10.1021/ja00221a015 38

[83] D. Bonchev. *Chemical Graph Theory: Introduction and Fundamentals*, volume 1. CRC Press, Boca Raton, FL, 1991. DOI: 10.1201/9781315139104 39

[84] A. T. Balaban. Applications of graph theory in chemistry. *Journal of Chemical Information and Computer Sciences*, 25(3):334, 1985. DOI: 10.1021/ci00047a033 39

[85] V. Sharma, C. Wang, R. G. Lorenzini, R. Ma, Q. Zhu, D. W. Sinkovits, G. Pilania, A. R. Oganov, S. Kumar, and G. A. Sotzing. Rational design of all organic polymer dielectrics. *Nature Communications*, 5:4845, 2014. DOI: 10.1038/ncomms5845 39

[86] C.-S. Liu, G. Pilania, C. Wang, and R. Ramprasad. How critical are the van der Waals interactions in polymer crystals? *Journal of Physical Chemistry A*, 116(37):9347, 2012. DOI: 10.1021/jp3005844 39

[87] G. Pilania and X.-Y. Liu. Machine learning properties of binary wurtzite superlattices. *Journal of Materials Science*, 53(9):6652, 2018. DOI: 10.1007/s10853-018-1987-z 39, 40

[88] W. Hume-Rothery and B. R. Coles. *Atomic Theory for Students of Metallurgy*. Maney Publishing, Leeds, England, 1988. DOI: 10.1088/0031-9112/11/11/015 41

[89] E. O. Hall. The deformation and ageing of mild steel: III Discussion of results. *Proc. Physical Society B*, 64(9):747, 1951. DOI: 10.1088/0370-1301/64/9/303 41

[90] N. J. Petch. The influence of grain boundary carbide and grain size on the cleavage strength and impact transition temperature of steel. *Acta Metallurgica*, 34(7):1387, 1986. DOI: 10.1016/0001-6160(86)90026-x 41

[91] O. Muller and R. Roy. *The Major Ternary Structural Families*. Springer, Berlin, 1974. 41

[92] G. Pilania, P. V. Balachandran, C. Kim, and T. Lookman. Finding new perovskite halides via machine learning. *Frontiers in Materials*, 3:19, 2016. DOI: 10.3389/fmats.2016.00019 41

[93] R. Ouyang, S. Curtarolo, E. Ahmetcik, M. Scheffler, and L. M. Ghiringhelli. SISSO: A compressed-sensing method for identifying the best low-dimensional descriptor in an immensity of offered candidates. *Physical Review Materials*, 2(8):083802, 2018. DOI: 10.1103/physrevmaterials.2.083802 42, 43, 46, 47

[94] R. Ouyang, E. Ahmetcik, C. Carbogno, M. Scheffler, and L. M. Ghiringhelli. Simultaneous learning of several materials properties from incomplete databases with multi-task SISSO. *Journal of Physics: Materials*, 2019. DOI: 10.1088/2515-7639/ab077b 42, 43

[95] R. Tibshirani, M. Wainwright, and T. Hastie. *Statistical Learning with Sparsity: The Lasso and Generalizations*. Chapman and Hall/CRC, London, 2015. DOI: 10.1201/b18401 42

[96] B. R. Goldsmith, M. Boley, J. Vreeken, M. Scheffler, and L. M. Ghiringhelli. Uncovering structure-property relationships of materials by subgroup discovery. *New Journal of Physics*, 19(1):013031, 2017. DOI: 10.1088/1367-2630/aa57c2 43

[97] C. J. Bartel, C. Sutton, B. R. Goldsmith, R. Ouyang, C. B. Musgrave, L. M. Ghiringhelli, and M. Scheffler. New tolerance factor to predict the stability of perovskite oxides and halides. *Science Advances*, 5(2):eaav0693, 2019. DOI: 10.1126/sciadv.aav0693 43

[98] C. J. Bartel, S. L. Millican, A. M. Deml, J. R. Rumptz, W. Tumas, A. W. Weimer, S. Lany, V. Stevanović, C. B. Musgrave, and A. M. Holder. Physical descriptor for the Gibbs energy of inorganic crystalline solids and temperature-dependent materials chemistry. *Nature Communications*, 9(1):4168, 2018. DOI: 10.1038/s41467-018-06682-4 43

[99] C. M. Acosta, R. Ouyang, A. Fazzio, M. Scheffler, L. M. Ghiringhelli, and C. Carbogno. Analysis of topological transitions in two-dimensional materials by compressed sensing. *ArXiv Preprint ArXiv:1805.10950*, 2018. 43

[100] P. Villars. *Pearson's Crystal Data: Crystal Structure Database for Inorganic Compounds*. ASM International, Materials Park, OH, 2007. 43, 44

[101] A. O. Oliynyk, E. Antono, T. D. Sparks, L. Ghadbeigi, M. W. Gaultois, B. Meredig, and A. Mar. High-throughput machine-learning-driven synthesis of full-Heusler compounds. *Chemistry of Materials*, 28(20):7324, 2016. DOI: 10.1021/acs.chemmater.6b02724 43

[102] L. Breiman. Random forests. *Machine Learning*, 45:5, 2001. DOI: 10.1515/9783110941975 43

[103] P. Villars, H. Okamoto, and K. Cenzual. ASM alloy phase diagrams database. *ASM International*, Materials Park, OH, 2006. 44

[104] A. Seidl, A. Görling, P. Vogl, J. A. Majewski, and M. Levy. Generalized Kohn-Sham schemes and the band-gap problem. *Physical Review B*, 53(7):3764, 1996. DOI: 10.1103/physrevb.53.3764 44

[105] L. J. Sham and M. Schlüter. Density-functional theory of the energy gap. *Physical Review Letters*, 51(20):1888, 1983. DOI: 10.1103/physrevlett.51.1888 44

[106] W. Setyawan, R. M. Gaume, S. Lam, R. S. Feigelson, and S. Curtarolo. High-throughput combinatorial database of electronic band structures for inorganic scintillator materials. *ACS Combinatorial Science*, 13(4):382, 2011. DOI: 10.1021/co200012w 44

[107] A. Nilsson, L. G. M. Pettersson, and J. Norskov. *Chemical Bonding at Surfaces and Interfaces*. Elsevier, Amsterdam, 2011. DOI: 10.1016/B978-0-444-52837-7.X5001-1

[108] R. Armiento, B. Kozinsky, M. Fornari, and G. Ceder. Screening for high-performance piezoelectrics using high-throughput density functional theory. *Physical Review B*, 84(1):014103, 2011. DOI: 10.1103/physrevb.84.014103

[109] I. E. Castelli, T. Olsen, S. Datta, D. D. Landis, S. Dahl, K. S. Thygesen, and K. W. Jacobsen. Computational screening of perovskite metal oxides for optimal solar light capture. *Energy and Environmental Science*, 5(2):5814, 2012. DOI: 10.1039/c1ee02717d

[110] O. Madelung. *Semiconductors: Data Handbook*. Springer, Heidelberg, 2012. DOI: 10.1007/978-3-642-18865-7

[111] I. E. Castelli, J. M. García-Lastra, K. S. Thygesen, and K. W. Jacobsen. Bandgap calculations and trends of organometal halide perovskites. *APL Materials*, 2(8):081514, 2014. DOI: 10.1063/1.4893495

[112] A. K. Singh, K. Mathew, H. L. Zhuang, and R. G. Hennig. Computational screening of 2D materials for photocatalysis. *Journal of Physical Chemistry Letters*, 6(6):1087, 2015. DOI: 10.1021/jz502646d

[113] R. Gautier, X. Zhang, L. Hu, L. Yu, Y. Lin, T. O. L. Sunde, D. Chon, K. R. Poeppelmeier, and A. Zunger. Prediction and accelerated laboratory discovery of previously unknown 18-electron ABX compounds. *Nature Chemistry*, 7(4):308, 2015. DOI: 10.1038/nchem.2207

[114] F. A. Rasmussen and K. S. Thygesen. Computational 2D materials database: Electronic structure of transition-metal dichalcogenides and oxides. *Journal of Physical Chemistry C*, 119(23):13169, 2015. DOI: 10.1021/acs.jpcc.5b02950

[115] M. J. Weber. *Handbook of Optical Materials*. CRC Press, Boca Raton, FL, 2018. DOI: 10.1201/9781315219615 44

[116] P. Dey, J. Bible, S. Datta, S. Broderick, J. Jasinski, M. Sunkara, M. Menon, and K. Rajan. Informatics-aided bandgap engineering for solar materials. *Computational Materials Science*, 83:185, 2014. DOI: 10.1016/j.commatsci.2013.10.016 45

[117] J. Lee, A. Seko, K. Shitara, K. Nakayama, and I. Tanaka. Prediction model of band gap for inorganic compounds by combination of density functional theory calculations and machine learning techniques. *Physical Review B*, 93(11):115104, 2016. DOI: 10.1103/physrevb.93.115104 45, 46

[118] G. Pilania, J. E. Gubernatis, and T. Lookman. Multi-fidelity machine learning models for accurate bandgap predictions of solids. *Computational Materials Science*, 129:156, 2017. DOI: 10.1016/j.commatsci.2016.12.004 46

CHAPTER 3

Learning with Large Databases

From a point of view that is both complementary and supplementary, this chapter revisits some of the uses and opportunities for materials design and discovery just discussed in Section 2.2.3. The previous subsection's focus was mainly about macroscopic-scale feature generation and selection and illustrated the benefit that attention on both topics produces with examples where data was gathered or generated for specific classes for specific applications. In this chapter, we are more concerned with the predictive potential of machine learning when applied to very large, community accessible databases. We review the available large databases, discuss some machine learning work using these data aimed at defining general rules for the likelihood of substituting in known compounds one element for another and illustrate the potential difference in predictive potential of using ground state vs. observed data of material structure and properties. We also explain decision tree-based versions of the random forests and gradient tree boosting machine learning methods. We note that these methods return a measure of the relative importance of the features and the dependency of one feature with the others.

3.1 DATABASES

In the recent past, the materials science community, both experimental and computational, have devoted attention to building databases that permit sharing and widespread dissemination. This is partly motivated by the stringent data management policies enforced by the federal funding agencies and partly by the community awareness driven by advancements made in other science domains such as cancer research, life sciences, etc. As a result, there has been an exponential growth and renewed excitement within the community to share structured data. The Minerals, Metals & Materials Society's (TMS's), "Building a Materials Data Infrastructure" report defines "materials data" as those primarily "used in the discovery, design, development, and implementation of materials, materials processing, and materials innovations" [1]. Digital data is crucial for any informatics study and several efforts in the past, including the "Integrated Computational Materials Engineering (ICME): A Transformative Discipline for Improved Competitiveness and National Security" and the "Materials Genome Initiative" have underscored its impact in paving the path for a paradigm shift in the way we do everyday materials science and engineering. In 2013, the TMS and Materials Research Society conducted a survey to gain insights into the view of materials science and engineering as a community on the subjects of open research and big data that had about 650 responses from the materials experts (64% from North America, 17% from Europe, and 13% from Asia) [1]. Two results that shed light on the community

Top 3 reasons to sharing data on an open access basis

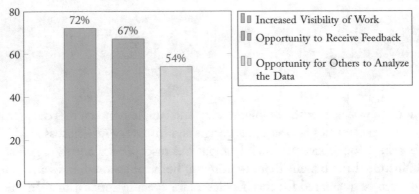

Top 3 impediments to sharing data on an open access basis

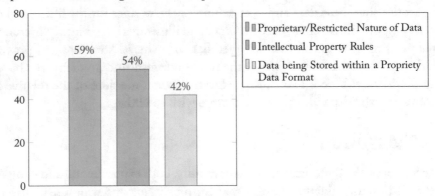

Figure 3.1: TMS and MRS survey results for two questions related to open data sharing [1].

viewpoint on open data sharing and access are shown in Figure 3.1 This is a topic area that has created substantial debate and will continue to do so in the near future.

The focus of this chapter is not to inform readers about the overall "philosophy" of materials database management standards, policies, and important challenges associated with it, but to provide a broad overview of some of the readily available databases to the general audience (open and closed that require subscription). We encourage the materials community to actively participate in policy discussions concerning data management and sharing. At the same time, we also encourage suggestions based on the lessons learned from related efforts in other scientific disciplines concerning some of the best and avoidable practices/policies.

On a much narrower sense of materials data for learning, design, and discovery, we can divide the field into three categories: (i) experimental, (ii) computational, and (iii) combined. Most of the numbers that are being reported have a timestamp that correspond to the time

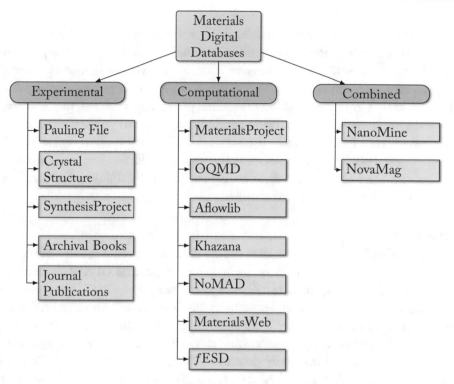

Figure 3.2: Classification of Materials Digital Databases based on how the source data. Broadly, we can classify the databases as experimental and computational. In addition, there is a growing trend to compile both in one repository (referred to as "Combined" here).

during which this book was written. A vast majority of the discussed databases are expected to grow organically as more data becomes available or generated. It is not possible to do justice to describe each database in one chapter because each database is an outcome of significant research efforts. The intended purpose is to provide a brief introduction to the readers about these infrastructures. We then encourage the readers to delve much deeper in to a specific resource that they see fit.

3.2   EXPERIMENTAL DATABASES

In this section, we focus on databases and repositories that only store legacy and current data from experimental outcomes.

Pauling File

This is one of the largest databases for inorganic materials that include data on phase diagrams, crystal structures, and physical properties. A more detailed information pertaining to the database can be accessed at `paulingfile.com`. This database contains 48,629 phase diagrams, 335,016 crystal structure entries, and 136,838 physical properties entries. The Pauling file is a relational database that groups crystallographic data, phase diagrams, and physical properties of crystalline inorganic materials. The data are curated from original publications, covering world scientific literature from 1891 to current literature. As of January 1, 2018, website indicates that about 106,421 journal publications have been processed to collect and organize the content. The Materials Platform for Data Science is an online edition of the Pauling file database and full access to the materials data requires subscription.

Crystal Structures

There are a host of online repositories that collate data on crystal structure of materials. One of the most popular databases known to us is the International Crystal Structure Database (ICSD)[2], which is accessed at `http://www2.fiz-karlsruhe.de/icsd_home.html`. The ICSD hosts more than 200,000 crystal structures and it is typically updated twice a year with data from scientific journals and other credible sources at the rate of 7,000 structures per year. ICSD contains data from both experimental and theoretical work, but we resort to this database to access information on experimentally determined crystal structures. According to the website, "The quality and completeness of the data is ensured by collaborating and networking with the Cambridge Crystal Data Center (CCDC) in Great Britain, the International Center for Diffraction Data in the USA, Technicum in Stuttgart and the Vinca Institute of Nuclear Science in Belgrade." Access to crystal structure data requires subscription.

SynthesisProject

The `synthesisproject.org` is an outcome of an innovative research project that uses natural language processing and text mining methods "to advance computational learning around materials synthesis approaches by creating a predictive synthesis system for advanced materials design and processing." Currently, the online portal has aggregated synthesis data on more than 30 different inorganic oxides systems. In addition to databases, the website also provides access to the machine learning models via the GitHub page.

Archival Books

Legacy archival books such as the Landolt–Bornstein compendium, Pearson crystallographic database, Springer handbook, Ternary Major Crystal Families, and handbooks still serve as a rich data source for materials informatics research. Some of the data in this book can also be found in the Pauling file (Section 3.2).

Journal Articles

In addition to the above, published articles in peer-reviewed journals remain the perennial source for materials informatics investigation. A vast majority of the digital repositories rely on the data and standards maintained in the publications.

3.3   COMPUTATIONAL DATABASES

In this section, we will focus on some of the online databases that are based on first principles-based DFT calculations. These repositories do not merely serve as a static resource for screening data and providing important information from first principles calculations. An important consequence is that these databases have enabled rapid exploration of materials properties in previously unexplored chemical and crystal structure spaces. A snapshot of some of the key properties stored by these computational databases is shown in Figure 3.3. Almost all of the computational databases are open-source and undergo regular updates. A majority of them are also equipped with machine learning capabilities for accomplishing specific basic tasks.

Materials Project

Materials Project uses high-throughput DFT as implemented in the Vienna Ab Initio Simulation Package (VASP) to calculate the total energy of compounds [3]. A vast majority of the crystal structure data that are used by the Materials Project are from the ICSD (Section 3.2). All calculations are performed at the Generalized Gradient Approximation (GGA) and GGA+U level. Typical data from this database include relaxed crystal structure, energy from the convex hull, elastic stiffness tensors, X-ray absorption spectra, piezoelectric tensors, calculated X-ray diffraction spectra, electronic band structure, and density of states spectra.

In addition, this database also contain several "apps" and each app, in turn, has specific functionalities that cater to addressing a specific problem. For instance, the phase diagram app allows the user to select elemental combinations and the output is a plot of Energy vs. Mole Fraction diagram that contains specific data points representing a given stoichiometry and the corresponding crystal structure. At the time of writing, there were a total of 14 apps. The database includes 133,691 inorganic compounds, 58,329 electronic band structures, 21,954 molecules, 530,243 nanoporous materials, 13,942 elastic tensors, 3,028 piezoelectric tensors, 3,628 intercalation electrodes, and 16,128 conversion electrodes. The website can be accessed at `www.materialsproject.org`.

Open Quantum Materials Database

Open Quantum Materials Database (OQMD) also uses high-throughput DFT-GGA($+U$) as implemented in VASP [4]. One of the main distinction between the OQMD and Materials Project is that the OQMD includes data on hypothetical compounds (in addition to those found in the ICSD). As a result, OQMD contains more data on inorganic compounds compared to that of the Materials Project. At the time of writing, OQMD had 563,247 entries. One dif-

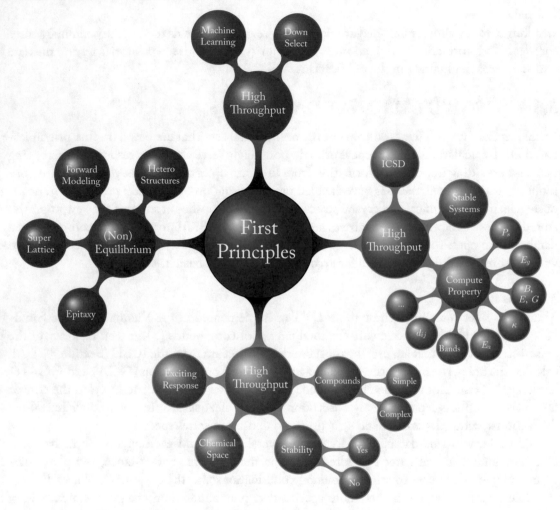

Figure 3.3: A partial list of some of the key materials properties stored in various computational databases.

ference between OQMD and the Materials Project is OQMD provides data on decomposition reactions, which allows one to evaluate the thermodynamic stability of arbitrary compounds. The database can be accessed at `http://oqmd.org`.

AFLOW

Aflow uses high-throughput DFT as implemented in VASP and Quantum Espresso to calculate physical properties such as relaxed geometries, electronic and phonon band structures, magnetic properties, and thermodynamic properties [5]. Aflow contains phonon band structures and phonon-related properties that are not found in the current implementation of Materials Project and OQMD. Similar to Materials Project and OQMD, it also features post-processing tools to automatically generate plots of electronic band structure and density of states, as well as plots of phase diagrams. The website can also be utilized to extract properties such as heat capacity, formation energy, and magnetic configuration. At the time of writing, the Aflow database contained 2,532,343 compounds and 341,866,170 calculated properties. The database can be accessed at `http://aflow.org`.

Khazana and Polymer Genome

Khazana is a computational database based on VASP that contains both inorganic and polymeric materials. The web platform Khazana also hosts an extensive database for "polymer genome," which is specifically geared toward expediting design and discovery of novel and improved polymers via utilizing experimental and computational data in combination with informatics based tools and techniques [6]. The database can be accessed at `https://khazana.gatech.edu`. The polymer genome project is available at `https://www.polymergenome.org`.

NoMaD

The Novel Materials Discovery (NoMaD) database maintains one of the largest repositories for input and output files of several computational materials science codes. The NoMaD repository contained 50,236,539 total energy calculations [7]. The database can be accessed at `https://nomad-coe.eu`.

MaterialsWeb

MaterialsWeb uses high-throughput DFT calculations using the $GGA(+U)$ method to develop a database of structure, electronic structure, and thermodynamic properties of 2D and 3D materials [8]. The database can be accessed at `https://materialsweb.org/`.

f-Electron Structure Database

f-ESD is an open access database dedicated to f-electron orbital systems which includes Lanthanides and Actinides as constituent elements [9]. This database uses Wien2k and DFT+Dynamical Mean Field Theory (DMFT) for capturing the strong electron correlation and spin-orbit coupling in the all-electron environment to yield a faithful representation for strongly correlated systems. This database contains 28,000 crystal structure files with approximately 9000 independent f-electron compounds. In addition, it also contains data on electronic band structures and density of states for approximately 3200 binary f-electron compounds. The database can be accessed at `http://correlatedmaterials-lanl.org`.

3.4   COMBINED DATABASES

In addition to the predominantly experimental and computational digital databases, there are additional databases that store data from both experiments and computations.

NanoMine

NanoMine (`https://qa.materialsmine.org`) stores data on polymer nano-composite processing, structure, and properties [10]. At the time of writing, the database hosts more than 500 results that include data on polymer composition, processing, structure, and properties (for example, glass transition temperature, AC dielectric dispersion, and dielectric breakdown strength). In addition to the data, the online portal also contain two tools: (i) module tools for statistical learning and analysis and (ii) simulation tools for materials property simulation using physics-based models. Several of the tools are under-construction.

NovaMag

This open database was built for designing novel "rare-earth free/lean permanent magnets" [11]. It includes both theoretical and experimental data about key magnetic material properties such as magnetic moments, magnetocrystalline anisotropy energy, exchange parameters, Curie temperature, domain wall width, exchange stiffness, coercivity, and maximum energy product. The database utilizes at least three computer simulation codes namely USPEX, VASP, and FP-LMTP code RSPt to accomplish the various structure relaxation and calculation of magnetic properties tasks. The website can be accessed at `http://crono.ubu.es/novamag/`.

Table 3.1 summarizes some of the major materials databases along with the uniform resource locator (URL).

3.5   MATERIALS DESIGN AND DISCOVERY

One of the overarching goals of building massive databases is to develop a principled approach that enables rapid design and discovery of new materials, which is in spirit with the materials genome initiative and integrated computational materials engineering. There are a number of published works that has demonstrated the potential of leveraging the data in these databases to design new materials. Some of the notable examples include design of battery materials and thermoelectric materials. This is a rapidly growing area in materials science and we anticipate more promising examples in the near future as the database grows in size and quality.

In fact, the idea of extracting trends and patterns from materials databases for materials design and discovery is not entirely new. In 1959, Mooser and Pearson were two of the earliest proponents of utilizing diagrammatic "structure sorting maps" for studying trends in solid-state materials using representations based on average principal quantum numbers and Pauling scale electronegativity differences [12]. Later, St. John and Bloch extended the approach to classify crystal structures of wide band gap octet AB compounds [13]. Since then Phillips [14–16],

Table 3.1: List of key data sources along with the URL and a category description based on the type of data contained in the database

Name	URL	Category
AFLOWLIB	https://aflowlib.org	Computational
ASM Alloy Center Database	https://mio.asminternational.org/ac	Alloys
Cambridge Crystallographic Data Centre	www.ccdc.cam.ac.uk	Crystallography
CatApp	https://suncat.stanford.edu/catapp	Catalysts
Citrination	https://citrination.com	General Materials Data
Computational Materials Repository	https://cmr.fysik.dtu.dk	Computational
Crystallography Open Database	http://www.crystallography.net	Crystallography
f-Electron Structure Database	https://correlatedmaterials–lanl.org	Computational
Harvard Clean Energy Project	cepdb.molecularspace.org	Computational
Inorganic Crystal Structure Database	https://icsd.fiz-karlsruhe.de/	Crystallography
Khazana	https://khazana.gatech.edu	Computational
Materials Project	https://materialsproject.org	Computational
MaterialsWeb	https://materialsweb.org	General Materials Data
MatNavi (NIMS)	https://mits.nims.go.jp/index en.html	General Materials Data
Nanomine	https://qa.materialsmine.org	Exp. and Computational
NIST Materials Data Repository	https://materialsdata.nist.gov	General Materials Data
NIST Interatomic Potentials Repository	www.ctcms.nist.gov/potentials	Computational
Novamag	https://crono.ubu.es/novamag	Exp. and Computational
NoMaD	https://nomad-coe.eu	Computational
Open KIM (Interatomic potentials)	https://openkim.org	Computational
Open Quantum Materials Database	https://oqmd.org	Computational
Pauling File	http://paulingfile.com	Experimental
Polymer Genome	https://www.polymergenome.org	Computational
SpringerMaterials	https://materials.springer.com	General Materials Data
Synthesis Project	https://synthesisproject.org	Experimental

Chelikowsky [17], Littlewood [18], Zunger [19, 20], Pettifor [21], Cohen [22], Andreoni [23], Burdett [24], Rabe [25], Villars [26], and others [27–30] have advanced the paradigm to classify structures and properties of many other binary and multicomponent crystalline compounds with varying degrees of accuracy and success. We touched briefly on this approach in Chapter 1.

More recently, data science and machine learning methods have emerged as a novel paradigm for extracting trends and patterns from small and large datasets [27, 31–43]. In these studies, data from both successful and failed experiments are used to train machine learning models, which in turn predict whether a new data sample (not present in the training set) can be discovered with properties better than those in the training data. Thus, these methods do not use or strictly require thermodynamic stability data but take advantage of the past experiments to make inference about future experiments. In addition, physics-based computations can augment the data-driven predictions by providing a theoretical basis for the predictive understanding and motivate new experiments. Numerous new compounds with desired structure and properties have been experimentally discovered with the aid of data-driven machine learning methods [44–51]. For the reminder of the chapter we will focus on two specific examples, namely substitution probability analysis and T = 0 K vs. T ≠ 0 K, to further motivate the key role played by these databases in aiding materials design and discovery.

3.5.1   SUBSTITUTION PROBABILITY ANALYSIS

The question of interest is "What is the likelihood that a chemical element A can be replaced by another element B in a given stoichiometry and what are the likely consequences of such substitution?" One of the key pathways to addressing this question is to define for a "physical or chemical similarity scale" (akin to electronegativity, hardness, Mendeleev number, and Pettifor scale to name a few) that captures the underlying physical meaning between the elements in the periodic table under some well-defined constraint. This definition is a non-trivial problem, and to-date there is no universal scale available to accomplish the task. This unavailability is mainly because of the complexities that arise due to the complex interplay between crystal structure and chemistry. Alternatively, one can also leverage the similarity metric that is intrinsic to a machine learning algorithm to address the question. In the literature, the common practice is to combine the physical or chemical scales of interest with supervised learning algorithms to accomplish the objective.

Hautier et al. [52] developed a compound and structure prediction method based on probabilistic principles. In their approach, random variables are defined that represent "what crystal structure is present at a given composition" for a compound. Let c_i represent a composition ABO_3, then the feature x_{c_i} will have values for the crystal structure type such as perovskite, ilmenite, aragonite, etc. Also, the specific condition "x_{c_i} = no structure" indicates the absence of any compound at the given stoichiometry. In addition, variables that represent the chemical constituents are also also developed (e.g., E_i =Ag, Cu, Na, etc.). Given these variables, one can define a vector $\vec{X} = (x_{c_1}, x_{c_2}, \cdots, x_{c_n}, x_{E_1}, x_{E_2}, \cdots, x_{E_n})$, where the composition space was discretized using n composition bins. A cumulant expansion truncated after the pair terms was defined to approximate the complex and multivariate probability function $p(\vec{X})$, which is defined as follows [53]:

$$p(\vec{X}) \approx \frac{1}{Z} \prod_i p(x_i) \prod_{j<k} \frac{p(x_i, x_j)}{p(x_i) p(x_j)}.$$

The parameters $p(x_i)$ and $p(x_i)p(x_j)$ are estimated from a database using a Bayesian approach (see Chapter 7 and Appendix A). In this work, Hautier et al. used the crystal structure and composition data stored in the ICSD for the substitutional analysis.

Balachandran et al. [54] developed a Bayesian approach to predict novel non-centro-symmetric (NCS) materials, i.e., materials that break the spatial inversion symmetry, in the $n = 1$ Ruddlesden–Popper (RP) oxides with stoichiometry A_2BO_4 (the crystal structure is shown in Figure 3.4). Such materials enable a number of technologically relevant macroscopic properties, including ferroelectricity, piezoelectricity, second-harmonic generation, and optical gyrotropy [55]. The key question is "Are there particular combinations of A and B cations that would stabilize a NCS phase?" This question can be addressed by applying Bayes' rules or Bayes' theorem (Appendix A), which is one of the well-known methods that combine prior experience, $p(\theta)$, with observed data or current evidence, $p(\mathcal{D})$, to make inference [*posterior*, $p(\theta|\mathcal{D})$] about the data [56, 57]. Mathematically, the theorem is expressed as

$$p(\theta|\mathcal{D}) = \frac{p(\mathcal{D}|\theta)p(\theta)}{p(\mathcal{D})}. \tag{3.1}$$

The prior, $p(\theta)$, is the strength of the belief in θ without the data $\mathcal{D}$ about chemical elements and/or temperature. The *posterior*, $p(\theta|\mathcal{D})$, is the strength in the belief θ when data $\mathcal{D}$ about chemical elements and/or temperature is/are given. In Equation (3.1), $p(\mathcal{D}|\theta)$ is usually referred to as the *likelihood*. The objective is to estimate the posterior, $p(\theta|\mathcal{D})$, from available structural data about A- and B-site elements (Figure 3.5), and temperature (Figure 3.6) to rapidly screen for A_2BO_4 chemical compositions and identify new and previously unexplored $n = 1$ RP chemistries that lift inversion symmetry. One can then use a structure optimization method to compute the relative stabilities of the predicted phases (for example, first principles calculations). The data for calculating the probabilities were taken from the journal publications.

How can you determine the posterior, $p(\theta|\mathcal{D})$? A simple worked out example is discussed here. The physical meaning for the various notations are not discussed here, and we encourage the interested readers to consult the original article. In this dataset, there were a total of 11 unique A- and 13 unique B-site elements. The probability of $X_2^+ \oplus X_3^+$ distortion, $p(\theta_{X_2^+ \oplus X_3^+}) = 18/89 = 0.2$, and the probability of P_4 distortion $p(\theta_{P_4}) = 7/89 = 0.078$. There are 22 occurrences of 214 RP oxides with Ca in the A site, obtained from the 18 occurrences with $X_2^+ \oplus X_3^+$ distortion, 4 occurrences with P_4 distortion, and no occurrences with other types of distortions. The overall probability of finding Ca in the A-site is $22/89 = 0.247$, i.e., $p(\mathcal{D}_{Ca}) = 0.247$. The likelihood of finding Ca in the A-site given the $X_2^+ \oplus X_3^+$ distortion is estimated as $p(\mathcal{D}_{Ca}|\theta_{X_2^+ \oplus X_3^+}) =$

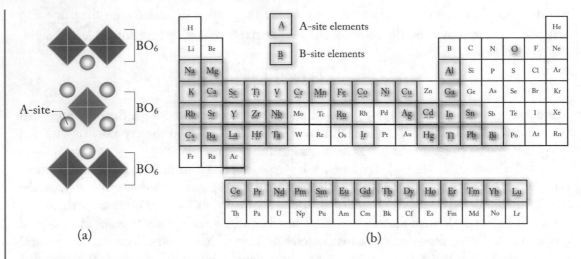

(a)   (b)

Figure 3.4: (a) The aristotype (parent crystal structure) of $n = 1$ Ruddlesden–Popper (RP) oxides in the $I4/mmm$ space group, where the crystallographic positions of A- and B-site elements are shown. (b) Periodic table of elements that highlight the A-site and B-site chemical elements that can potentially occupy the $n = 1$ RP structure. The search space is vast, and only a tiny fraction is experimentally explored. Substitution probability analysis can be used to glean insights about the role of chemical substitutions on the thermodynamic stability and ground state structure.

$18/18 = 1$. Similarly, the likelihood of observing Ca element in the A-site given the P_4 distortion is estimated as $p(\mathcal{D}_{Ca}|\theta_{P_4}) = 4/7 = 0.57$, and finally, the likelihood for any other distortion is zero. The posterior probabilities are obtained as follows:

$$p(\theta_{X_2^+ \oplus X_3^+}|\mathcal{D}_{Ca}) = \frac{1.0 \times 0.2}{0.247} = 0.81$$

$$p(\theta_{P_4}|\mathcal{D}_{Ca}) = \frac{0.57 \times 0.078}{0.247} = 0.19.$$

Bayes' rule captures key physical features associated with Ca element in the A-site of the 214 oxides: Among the BO_6 octahedra tilts modes, there is a relatively stronger tendency for the coupled $X_2^+ \oplus X_3^+$ rotation mode to occur than the P_4 tilt distortion. The propensity toward octahedral tilting could be rationalized based on the relatively small size of Ca^{2+} in the 214 lattice: simple BO_6 octahedral tilts are more likely to occur so as to produce more favorable bond-valences for the cations. The authors also computed the total energies of Ca_2IrO_4 in the $Pbca$, $Pca2_1$, $I4_1/acd$, and seven other symmetries using DFT. Centro-symmetric $Pbca$ and NCS $Pca2_1$ are found to be the lowest energy structures among all the candidate geometries— the parent phase is found to be extremely high in energy, indicating if the RP phase can be stable, the structure will be highly distorted.

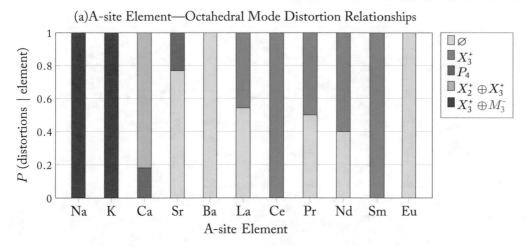

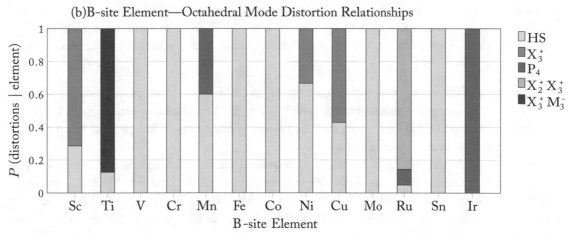

Figure 3.5: Posterior probabilities that a particular distortion in 214 RP oxides will be observed given a specific (a) *A*-site or (b) *B*-site element. These plots make it possible to rationally select *A*- and *B*-site elements conducive to specific structural distortions in 214 RP oxides. The fact that we have nonzero probabilities indicate that there is a strong interaction between *A*-, *B*-site elements and temperature, which Balachandran et al. do not deconvolve in the paper.

3.5.2 T = 0 vs. T ≠ 0 K

In DFT calculations, one of the key quantities of interest is the T = 0 K ground state total energy. However, real materials are synthesized and characterized at T ≠ 0 K. As a result, there is a gap when we seek to link the data generated from DFT computations with physical experiments. Typically, the total energy difference between two phases will provide a zeroth-order approximation to infer about the relative stability of a compound. However, there have been

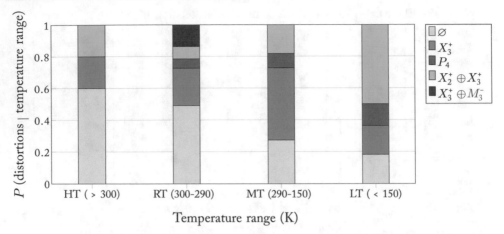

Figure 3.6: Posterior probabilities that a particular distortion in 214 RP oxides will be observed as a function of temperature or a temperature range. HT, RT, MT, and LT stands for high temperature, room temperature, intermediate temperature, and low temperature, respectively.

advancements in the recent past to overcome the gap and provide insights to guide new experiments. One of the specific developments, which is the main focus of this section, is in the realm of using convex hull (CH) analysis to address materials discovery challenges that can guide the design of metastable compounds [58].

As remarked earlier, DFT calculations provide a means to identify stable compounds. A particular implementation is based on "high-throughput calculations" in which a large series of DFT calculations are evaluated in an automatic high-throughput (HT) fashion. The total energy data for the compounds in the HT-DFT dataset can then be used to perform a convex hull analysis in which the lowest energy combination of phases can be identified for an arbitrary stoichiometry (for example, ABO_3) and the crystal structure-type that leads to the lowest energy from a group of pre-selected competing phases is identified. The pre-selected competing phases include (i) alternative crystal structures of the target compound and (ii) decomposition products of the target compound (such as $A+B+3/2O_2$, or $AO+BO_2$). A target compound with an energy lower than those in the combined groups (i) and (ii) is declared a stable ground state structure. If its energy is higher, then it is declared unstable. If unstable, but close to the convex hull, the compound might still be potentially formable as a metastable phase with a reasonable level of degree of metastability. A number of open source implementations of the DFT convex hull approach to predicting compound stability are available [3–5]. Different open sources databases can give different predictions depending on the details of the DFT and the construction of the convex hull. Although a convex hull analysis will provide insights about metastability of a compound, the exact temperature and pressure conditions to realize the metastable phase cannot be predicted.

These various points are illustrated in the work of Legrain et al. [59] and Balachandran et al. [60] and can have important implications. Legrain et al. were interested in the discovery on new half-heuslers, while Balachandran et al., were interested in new ABO_3 perovskites. Both groups combed the ICSD and recent literature for known formable compounds of the right chemical compositions. They then divided these into two groups—half-heusler or not or perovskite or not; that is, they defined the discovery problem to be one of binary classification. "Not" was something not in the proper space group or perhaps something that decomposed. They then specified a set of features, choose a machine learning classifier, and cross-validated the data. The result was a model into which features of prototype compounds are inputed and outputted are the predictions of being a half-heusler (or a perovskite) or something else which is known only to be not a half-heusler (or a perovskite). The questions are how well do models built with the ICSD data reproduce the ICSD data, how well do models built with the DFT data and CH analysis do the same, and how consistent are both approaches relative to their predictions of prototypes. The two groups found the *ab initio* predictions from the CH analysis of databases were significantly inconsistent with the experimentally observed structures and often with each other, but machine learning models trained on the experimental data had cross-validated predictions of accuracy of at least 90% in predicting the known compounds.

Balachandran et al. [60] emphasized that part of the reason for inconsistencies is that the CH analysis and machine learning predict two different substantive quantities. The CH analysis predicts the zero temperature mechanical stability of a crystal structure; the machine learning models predict formability of a structure because they are trained on data of materials formed. If a degree of metastability is added to the CH analysis, as suggested by Sun et al. [58], that admits materials whose energies within a certain distances above the convex hull, Balachandran et al. found the OQMD stability predictions show more agreement with compounds known to have been formed. Still, this agreement is only 67%. They proposed that if machine learning predicts a compound is formable and OQMD predicts it is stable, then this compound is more likely to be formable experimentally than other machine learning predicted prototypes.

For the binary classifiers, Legrain et al. used random forests [61], and Balachandran et al. used random forests and gradient tree boosting [62]. These machine learning methods are examples of ensemble methods, meaning that their results are combinations of the results of more than one model. We discuss these methods in the next section.

In addition to the convex hull analysis, data-driven approaches have also been explored to directly predict Gibbs free energy of inorganic solids for making inference about the stability of materials at elevated temperatures [63]. This approach relies on both experimental Gibbs free energy data and DFT computed features (for example, from the Materials Project database) for training the models. Since temperature is an input to the model, this approach has the potential to predict thermodynamic stability at $T \neq 0$ K. More traditional approaches to predicting thermodynamic stability involve quasi-harmonic approximation within the DFT formalism and thermo-calc (CALPHAD). Since these approaches do not readily make use of the databases (discussed earlier in the chapter), we do not discuss them here.

3.6   ENSEMBLE METHODS BASED ON DECISION TREES

We now discuss the random forests and gradient tree boosting methods. These methods were used in the T $= 0$ vs. T $\neq 0$ K work just discussed. While they are also used for regression analysis, we will only discuss their use for classification. They are *ensemble methods* [64–66] whose predictions are the votes or averages over those from many machine learning analyses of weighted or resampled adaptions of the training data. By combining the results of a number of sightly different methods, the expectation is the prediction of the combination is most likely more stable and has less variance from the true result than the prediction from just one method. The key characteristic is using combinations of methods that are diverse in the sense that that do not always agree in their predictions for the same single training data. While some ensemble methods might combine the predictions from quite dissimilar methods, random forests and gradient tree boosting both use decision trees as the core method [67].

Random forests is based on the concept of bagging; *gradient tree boosting* is based on boosting. The goal of either approach is to address the inherent machine learning problem of bias vs. variance control. In general, low bias models have high variance and vice versa. *Bagging*, which typically uses strong (small bias) learners, is mainly a variance reduction technique; *boosting*, which typically uses weak learners (high bias)—methods whose accuracies are slightly better that random guessing—is mainly a bias reduction technique [68].

Besides boosting and bagging, *stacking* [65] is sometimes used to create an ensemble method. Stacking typically combines models of different types. The different types might learn well some part of the problem. The idea to create an ensemble that will learn the entire problem. We will not discuss it further.

3.6.1   DECISION TREES

We start by describing the *decision tree method* for classification problems [64–66]. Suppose we have a training set of n observations $\mathcal{D} = \{(\boldsymbol{x}_i, c_i), i = 1, \ldots, N\}$ where an $\boldsymbol{x}_i$ is an input vector of D features and the c_i have the value of one of the K classes associated with this data. The most common case is that of *binary classification* where $K = 2$. Here, the c_i may have the values of True or False, Yes or No, etc., that map onto two integers, for example, 1 and 0. The analysis we present applies to the binary case as well as to cases where $K > 2$.

The method starts by finding a feature and a condition on the feature for the root node of the tree that splits the data into K leaves which have the least impurity. *Impurity* is some measure of the degree to which the spilt segregates the data into the separate classes. To establish the root, we examine each feature in turn, then according to the values of that feature in the data, we split the data and compute the impurity of the split. The feature and its value that has the smallest impurity of all the splits is the condition for the root node.

Three common impurity functions are the minority class, Gini index, and entropy [65]. The *minority class* measures the proportion of misclassified examples if the leaf was labeled with the majority. The *Gini index* is the expected error if we label the examples in the leaf randomly. The *entropy* is the expected information you gain if you know the class of a randomly drawn example. If the *impurity* of leaf $\mathcal{D}_\ell$ is $\mathrm{Imp}(\mathcal{D}_\ell)$ and $\mathcal{D}_1, \mathcal{D}_2, \ldots, \mathcal{D}_\ell$, where $\ell \leq K$, are the mutually exclusive leaves, then the impurity of the split is the weighted average

$$\mathrm{Imp}(\mathcal{D}_1, \ldots, \mathcal{D}_\ell) = \sum_{j=1}^{\ell} \frac{|\mathcal{D}_j|}{|\mathcal{D}|} \mathrm{Imp}(\mathcal{D}_j).$$

After the root split is established, we then repeat the process for each leaf. Thus, the tree is built recursively. The question is when to stop. If a leaf is pure or empty, we cease branching from it. How we stop at other leaves varies with implementations. We might stop if the amount of data associated with a leaf falls below a threshold, we might simply stop at a certain depth of the tree, or we might limit the number of leaves. Usually, the class of the data assigned to the leaf is based on a majority rule.

Once we stop, we have a model that allows us to classify a proposed new data point. To do this, we simply trace a path through the nodes based on the values of the features of the data point until it terminates at a leaf. The class assigned to the leaf determines the class assigned to the new data point.

The decision tree for classification is a simple and popular method that with slight modifications is also useful for regression, ranking, clustering and probability density estimation. The depth of the tree determines the tightness of the fit: The more nodes needing transversal to a leaf the more constraints a feature has to satisfy to arrive at the leaf defining its class. Slight variations in the data can lead to quite different nodal structures.

3.6.2 BAGGING AND RANDOM FORESTS

Bagging refers to ensemble methods that create diverse models from different random samples of the original data [64–66]. *Bootstrapping* is used to select these samples randomly and independently which means they are selected with replacement. A model $M_m(\mathbf{x})$ is built for each bootstrapped sample. The equally weighted collection of M of these models constitutes the ensemble model

$$\mathcal{M}(\mathbf{x}) = \frac{1}{M} \sum_{m=1}^{M} M_m(\mathbf{x}).$$

It is straightforward to use decisions trees within bagging. Bagging is a highly parallelizible method.

The method of *random forests* refers to a bagging construction but with an additional twist: For a given bootstrapped sample, it chooses at each node a subset of the features randomly and then branches. The process is repeated until building the tree is stopped. Each tree built uses subsets of the same size. The resampling of the data and random subset selection of the features add diversity to the predictions. The variance of the predictions of a random forest is reduced by increasing the size of the ensemble.

3.6.3   BOOSTING AND GRADIENT TREE BOOSTING

Boosting is a more involved class of ensemble methods [64–66]. Instead of adding independently learned members, it learns them sequentially. It starts by assigning each member of the data the same value of a weight. After classification, if the error rate is less than a half, it reassigns the weights by assigning the portion of the data misclassified half the total weight, thereby increasing (boosting) their weights, and the rest of the weight is assigned to the data correctly classified, thereby decreasing their weights. A confidence factor is then computed that determines the weight of the just constructed model in the entire ensemble. With repetitions of the procedure, the successive learners are forced to concentrate on the examples missed by the previous learner. Boosting is well suited for classification problems.

In a bit more detail, if $|\mathcal{D}|$ is the number of data, then each member of the data set starts with the weight $1/|\mathcal{D}|$. The classifier is run on the data to produce a model M_1. If the misclassification error rate is ϵ_1, then if $\epsilon_1 \geq 1/2$, the procedure is stopped. If not, the weights of the misclassified data are multiplied by $1/2\epsilon_1$, and those of the correctly classified data by $1/2(1 - \epsilon_1)$. These rescalings ensure the total weight still sums to one. A confidence factor (weight) α_1 is now assigned to the model, with a common choice being $\alpha_1 = \ln \sqrt{\frac{1-\epsilon_1}{\epsilon_1}}$. The steps are repeated until the desired number of models M is created. The ensemble model is

$$\mathcal{M}(\mathbf{x}) = \sum_{m=1}^{M} \alpha_m M_m(\mathbf{x}).$$

When used with decision trees, the tree depth is kept shallow, often only having only the root node, so that each tree constructed is a weak classifier, that is, one whose predictions are only slightly better than guessing randomly. A tree with only the root node is called a *stump*.

Gradient boosting adds the gradient decent (steepest decent) method to boosting as a way to decrease the bias as the classifier is built step by step. It builds a classifier by using a regressor. The method does not assign weights to the data, and its procedure for assigning weights to members of the ensemble is more involved that basic boosting, but the result is powerful. The *gradient decent* method is an iterative one that determines the parameters that optimize some objective function. The gradient boosting method builds the additive model $\mathcal{M}(\mathbf{x})$ by executing the steps

$$\mathcal{M}_m(\mathbf{x}) = \mathcal{M}_{m-1}(\mathbf{x}) + \alpha_m M_m(\mathbf{x}). \tag{3.2}$$

Here, $M_m(\mathbf{x})$ is the base regressor of the ensemble at step m. At each step, its parameters are chosen to minimize a loss function L given the current model $\mathcal{M}_{m-1}$ and its fit to the data

$$\mathcal{M}_m(\mathbf{x}) = \mathcal{M}_{m-1}(\mathbf{x}) + \arg\min_M \sum_{i=1}^{N} L(c_i, \mathcal{M}_{m-1}(\mathbf{x}_i) + M(\mathbf{x})). \tag{3.3}$$

The minimization is accomplished by gradient decent for which we have

$$\mathcal{M}_m = \mathcal{M}_{m-1} - \alpha_m \sum_{i=1}^{N} \frac{\partial L(c_i, \mathcal{M}_{m-1}(\mathbf{x}_i))}{\partial \mathcal{M}_{m-1}(\mathbf{x}_i)}, \tag{3.4}$$

where

$$\alpha_m = \arg\min_\alpha \sum_{i=1}^{N} L\left(c_i, M_{m-1}(\mathbf{x}_i) - \alpha \frac{\partial L(c_i, \mathcal{M}_{m-1}(\mathbf{x}_i))}{\partial \mathcal{M}_{m-1}(\mathbf{x}_i)}\right).$$

By comparing (3.2) and (3.4), we see that $M(\mathbf{x})$ is a function fitted to the negative gradient of the loss function. Hence, it is a regressor as opposed to a classifier.

Algorithms for gradient boosting regression and classification differ mainly in the types of loss functions used. One advantage of the method, whether it is for regression or classification, is its having several choices for the loss function. For a K class classification problem, a common choice of loss function is the multinomial deviance

$$L(c_i, M(\mathbf{x}_i)) = -\sum_{k=1}^{K} c_i^k \log \pi^k(\mathbf{x}_i), \tag{3.5}$$

where $c_i^k = 1$ if $c_i = k$ and 0 otherwise, which provides a probability for the estimates,

$$\pi^k(\mathbf{x}_i) = \frac{e^{M^k(\mathbf{x}_i)}}{\sum_{k=1}^{K} e^{M^k(\mathbf{x}_i)}}. \tag{3.6}$$

At the mth step, this choice leads to the following expression for the gradient:

$$\frac{\partial L(c_i^k, \mathcal{M}_{m-1}(\mathbf{x}_i))}{\partial \mathcal{M}_{m-1}(\mathbf{x}_i)} = c_i^k - \pi_m^k(\mathbf{x}_i), \tag{3.7}$$

where the dummy variable $c_i^k = 1$ if $c_i = k$ and 0 otherwise. Implied by (3.5) and (3.6) is the base regressor performs an independent fit to the gradient of the loss for each class value. With introduction of the dummy variable, in essence what is happening is at each iterations step, the original dataset $\mathcal{D} = \{(\mathbf{x}_i, c_i), i = 1, \ldots, N\}$ is replaced by K datasets $\mathcal{D}^k = \{(\mathbf{x}_i, c_i^k - \pi_m^k(\mathbf{x}_i)), i = 1, \ldots, N\}$ for which a regression model is built for each.

In *gradient tree boosting*, our core regressor is a tree regressor. The basics of the algorithm are: At each step of the iteration, we find the negative values of the K gradients (3.7), fit a regression tree $M_m^k(\mathbf{x})$ to each, and then update to the next step for each class via

$$\mathcal{M}_m^k = \mathcal{M}_{m-1}^k + \alpha_m^k M_m^k.$$

It is important to note again that even though we are building a classifier, we use a regression tree to connect the gradient and the data. On convergence, $\hat{\pi}^k(\mathbf{x})$ (3.6) is the probability the feature $\mathbf{x}_i$ belongs to class k. The predictor for the class value of some data $\mathbf{x}$ not in the training set is the class with the highest probability,

$$\hat{c} = \arg\max_k \hat{\pi}^k(\mathbf{x}).$$

An important modification to the above is changing the iteration (3.2) to

$$\mathcal{M}_m(\mathbf{x}) = \mathcal{M}_{m-1}(\mathbf{x}) + \mu\alpha_m M_m(\mathbf{x}),$$

μ has value between 0 and 1 and is called the *learning rate* because it scales the step size of the gradient decent procedure to make the procedure more stable by making the decent more gradual and smooth. Small learning rates require more members of the ensemble.

Stochastic gradient tree boosting is another embellishment of the method. Roughly speaking, it adds additional diversity to the procedure by using a random forest regressor in the sense that subsampling of the data and feature subsets are used. Another advantage of using these techniques is the reduction of computation time. While powerful, gradient tree boosting can become computationally intensive for large datasets that have large numbers of features. Fortunately, a number of implementations of the method are available for downloading. They each do things slightly differently.

3.6.4   CROSS-VALIDATION, RELATIVE FEATURE IMPORTANCE, AND PARTIAL DEPENDENCIES

Cross-Validation

Validating the models built by the methods just described is usually done by the standard machine-learning technique of *cross-validation*; that is to say, we build the model on a subset of the data, called the training data, and then adjust the parameters of the model to give good predictions for the held out data, called the testing data. Numerous cross-validation schemes exist and are discussed in standard texts [64–66].

The basic parameters of a decision tree implementation requiring adjustment include the depth of the tree and the minimum data allowed in a leaf. For random forests, adjusting the number of bootstrapped samples and the size of the features subset are additionally required.

Gradient tree boosting inherits the adjustable parameters of the decision tree method. In addition, care is needed in adjusting the learning rate and the size of the ensemble. For stochastic gradient tree boosting, depending on the implementation, additional adjustable parameters might include the sizes of the subsampling and feature subset used.

Random forests and in particular stochastic gradient tree boosting are typically computationally more intensive than other machine learning methods for classification. They are, however, quite powerful and are balanced counter-points with respect to the need to adjust bias and control variance. Their software implementations are readily available and well documented in such freely available packages such as scikit-learn [69] and R [70].

Relative Feature Importance

The proper choice of features is an important step in building effective models. In general, we can always improve our model's fit to the data by simply choosing more and more features. Besides the potential of overfitting, obtaining a good fit this way will generally not increase the understanding of the relation of the features and the material's properties and behaviors we are studying. Decision trees, however, have a relatively natural and simple empirical way of assessing the relative importances of the features we are using [64]. We can then exclude those that seem unimportant and then compare the performance of the model with and without them as a test. This capability assists learning from the data.

Different implementations seem to use different definitions of *relative feature importance*, but they are all based on the observations the some features split more data than the others or they split the data more frequently than the others. In either case, this suggests these features are relatively more important than the others. This information is easily gathered and leads to good a first-cut at feature reduction and selection.

Partial Dependencies

Gradient boosting methods return a joint probability $P(\mathbf{x}) = P(x_1, x_2, \ldots, x_D)$ for the values of the features x_i. This probability allows the computation of

$$F(\mathbf{x}) = \log \frac{P(\mathbf{x})}{1 - P(\mathbf{x})},$$

which is the logarithm of the *odds*. With a tree algorithm, it is possible to integrate out select features empirically and compute, for example,

$$F(x_1) = \int F(x_1, x_2, \ldots, x_d) dx_2 dx_3 \cdots dx_D,$$

or

$$F(x_1, x_2) = \int F(x_1, x_2, \ldots, x_d) dx_3 \cdots dx_D.$$

With the latter it becomes possible to compare $F(x_1, x_2)$, $F(x_1, x_3)$, $F(x_1, x_4)$, etc., to explore how one feature pairs with another to influence predictions. The marginalizations of $F(x_1, x_2, \ldots, x_D)$ are called *partial dependencies* [64]. Using this capability also assists learning from the data.

3.7   REFERENCES

[1] The Minerals Metals and Materials Society (TMS). *Building Materials Data Infrastructure: Opening New Pathways to Discovery and Innovation in Science and Engineering*. TMS, Pittsburgh, PA, 2017. 59, 60

[2] A. Belsky, M. Hellenbrandt, V. L. Karen, and P. Luksch. New developments in the inorganic crystal structure database (ICSD): Accessibility in support of materials research and design. *Acta Crystallographica B*, 58:364, 2002. DOI: 10.1107/s0108768102006948 62

[3] A. Jain, S. P. Ong, G. Hautier, W. Chen, W. D. Richards, S. Dacek, S. Cholia, D. Gunter, D. Skinner, G. Ceder, and K. A. Persson. Commentary: The materials project: A materials genome approach to accelerating materials innovation. *APL Materials*, 1(1), 2013. DOI: 10.1063/1.4812323 63, 72

[4] J. E. Saal, S. Kirklin, M. Aykol, B. Meredig, and C. Wolverton. Materials design and discovery with high-throughput density functional theory: The Open Quantum Materials Database (OQMD). *JOM*, 65(11):1501, 2013. DOI: 10.1007/s11837-013-0755-4 63

[5] S. Curtarolo, W. Setyawan, S. Wang, J. Xue, K. Yang, R. H. Taylor, L. J. Nelson, G. L. W. Hart, S. Sanvito, M. Buongiorno-Nardelli, N. Mingo, and O. Levy. AFLOWLIB.ORG: A distributed materials properties repository from high-throughput ab initio calculations. *Computational Materials Science*, 58:227, 2012. DOI: 10.1016/j.commatsci.2012.02.002 64, 72

[6] T. D. Huan, A. Mannodi-Kanakkithodi, C. Kim, V. Sharma, G. Pilania, and R. Ramprasad. A polymer dataset for accelerated property prediction and design. *Science Data*, 3:160012, 03 2016. DOI: 10.1038/sdata.2016.12 65

[7] C. Draxl and M. Scheffler. NOMAD: The FAIR concept for big data-driven materials science. *MRS Bulletin*, 43(9):676–682, 2018. DOI: 10.1557/mrs.2018.208 65

[8] K. Mathew, A. K. Singh, J. J. Gabriel, K. Choudhary, S. B. Sinnott, A. V. Davydov, F. Tavazza, and R. G. Hennig. MPInterfaces: A materials project based Python tool for high-throughput computational screening of interfacial systems. *Computational Materials Science*, 122:183, 2016. DOI: 10.1016/j.commatsci.2016.05.020 65

[9] H. Hafiz, A. I. Khair, H. Choi, A. Mueen, A. Bansil, S. Eidenbenz, J. Wills, J.-X. Zhu, A. V. Balatsky, and T. Ahmed. A high-throughput data analysis and materials discovery tool for strongly correlated materials. *NPJ Computational Materials*, 4(1):63, 2018. DOI: 10.1038/s41524-018-0120-9 65

[10] H. Zhao, X. Li, Y. Zhang, L. S. Schadler, W. Chen, and L. C. Brinson. Perspective: NanoMine: A material genome approach for polymer nanocomposites analysis and design. *APL Materials*, 4(5):053204, 2016. DOI: 10.1063/1.4943679 66

[11] P. Nieves, S. Arapan, J. Maudes, R. Marticorena, N. L. Del Brío, A. Kovacs, C. Echevarria-Bonet, D. Salazar, J. Weischenberg, H. Zhang, et al. Database of novel magnetic materials for high-performance permanent magnet development. *ArXiv:1902.05241*, 2019. DOI: 10.1016/j.commatsci.2019.06.007 66

[12] E. Mooser and W. B. Pearson. On the crystal chemistry of normal valence compounds. *Acta Crystallographica*, 12:1015, 1959. DOI: 10.1107/s0365110x59002857 66

[13] J. St. John and A. N. Bloch. Quantum-defect electronegativity scale for nontransition elements. *Physical Review Letters*, 33:1095, 1974. DOI: 10.1103/physrevlett.33.1095 66

[14] J. C. Phillips and J. A. Van Vechten. Spectroscopic analysis of cohesive energies and heats of formation of tetrahedrally coordinated semiconductors. *Physical Review B*, 2:2147, September 1970. DOI: 10.1103/physrevb.2.2147 66

[15] J. C. Phillips. Structural pseudoion form factors. *Solid State Communications*, 22(9):549–550, 1977. DOI: 10.1016/0038-1098(77)90132-6

[16] E. S. Machlin, T. P. Chow, and J. C. Phillips. Structural stability of suboctet simple binary compounds. *Physical Review Letters*, 38:1292, 1977. DOI: 10.1103/physrevlett.38.1292 66

[17] J. R. Chelikowsky and J. C. Phillips. Quantum-defect theory of heats of formation and structural transition energies of liquid and solid simple metal alloys and compounds. *Physical Review B*, 17:2453, 1978. DOI: 10.1103/physrevb.17.2453 68

[18] P. B. Littlewood. Structure and bonding in narrow gap semiconductors. *Critical Reviews in Solid State and Materials Sciences*, 11(3):229, 1983. DOI: 10.1080/01611598308244064 68

[19] A. Zunger. Systematization of the stable crystal structure of all AB-type binary compounds: A pseudopotential orbital-radii approach. *Physical Review B*, 22:5839, 1980. DOI: 10.1103/physrevb.22.5839 68

[20] T. R. Paudel, A. Zakutayev, S. Lany, M. d'Avezac, and A. Zunger. Doping rules and doping prototypes in A_2BO_4 spinel oxides. *Advanced Functional Materials*, 21(23):4493, 2011. DOI: 10.1002/adfm.201101469 68

[21] D. G. Pettifor. Structure maps revisited. *Journal of Physics: Condensed Matter*, 15(25):V13, 2003. DOI: 10.1088/0953-8984/15/25/402 68

[22] M. L. Cohen. Electronic charge densities in semiconductors: Electron density calculations give new insights into the origins of the properties of solids. *Science*, 179(4079):1189, 1973. DOI: 10.1126/science.179.4079.1189 68

[23] W. Andreoni and G. Galli. Unified structural classification of AB_2 molecules and solids from valence electron orbital radii. *Physics and Chemistry of Minerals*, 14(5):389, 1987. DOI: 10.1007/bf00628814 68

[24] J. K. Burdett and S. L. Price. An interpretation of structural sorting diagrams for AB type compounds using molecular orbital ideas. *Journal of Physics and Chemistry of Solids*, 43(6):521, 1982. DOI: 10.1016/0038-1098(82)90166-1 68

[25] K. M. Rabe. Quantum diagrams and prediction of new materials. *Journal of Alloys and Compounds*, 197:131, 1993. DOI: 10.1016/0925-8388(93)90035-1 68

[26] P. Villars. Three-dimensional structural stability diagrams for 648 binary AB_3 and 389 binary A_3B_5 intermetallic compounds: III. *Journal Less Common Metals*, 102(2):199, 1984. DOI: 10.1016/0022-5088(84)90316-3 68

[27] K. Rajan. Materials informatics: The materials "gene" and big data. *Annual Review of Materials Research*, 45(1):153, 2015. DOI: 10.1146/annurev-matsci-070214-021132 68

[28] D. Lencer, M. Salinga, B. Grabowski, T. Hickel, J. Neugebauer, and M. Wuttig. A map for phase-change materials. *Nature Materials*, 7:972, 2008. DOI: 10.1038/nmat2330

[29] Y. Saad, D. Gao, T. Ngo, S. Bobbitt, J. R. Chelikowsky, and W. Andreoni. Data mining for materials: Computational experiments with AB compounds. *Physical Review B*, 85:104104, 2012. DOI: 10.1103/physrevb.85.104104

[30] A. Seko, T. Maekawa, K. Tsuda, and I. Tanaka. Machine learning with systematic density-functional theory calculations: Application to melting temperatures of single- and binary-component solids. *Physical Review B*, 89:054303, 2014. DOI: 10.1103/physrevb.89.054303 68

[31] A. O. Oliynyk, E. Antono, T. D. Sparks, L. Ghadbeigi, M. W. Gaultois, B. Meredig, and A. Mar. High-throughput machine-learning-driven synthesis of full-Heusler compounds. *Chemistry of Materials*, 28(20):7324, 2016. DOI: 10.1021/acs.chemmater.6b02724 68

[32] T. K. Patra, V. Meenakshisundaram, J.-H. Hung, and D. S. Simmons. Neural-network-biased genetic algorithms for materials design: Evolutionary algorithms that learn. *ACS Combinatorial Science*, 19(2):96, 2017. DOI: 10.1021/acscombsci.6b00136

[33] T. Ueno, T. D. Rhone, Z. Hou, T. Mizoguchi, and K. Tsuda. COMBO: An efficient Bayesian optimization library for materials science. *Materials Discovery*, 4:18, 2016. DOI: 10.1016/j.md.2016.04.001

[34] P. V. Balachandran, D. Xue, J. Theiler, J. Hogden, and T. Lookman. Adaptive strategies for materials design using uncertainties. *Scientific Reports*, 6:19660, 01 2016. DOI: 10.1038/srep19660

[35] L. M. Ghiringhelli, J. Vybiral, S. V. Levchenko, C. Draxl, and M. Scheffler. Big data of materials science: Critical role of the descriptor. *Physical Review Letters*, 114:105503, 2015. DOI: 10.1103/physrevlett.114.105503

[36] G. Pilania, J. E. Gubernatis, and T. Lookman. Structure classification and melting temperature prediction in octet AB solids via machine learning. *Physical Review B*, 91:214302, 2015. DOI: 10.1103/physrevb.91.214302

[37] G. Pilania, C. Wang, X. Jiang, S. Rajasekaran, and R. Ramprasad. Accelerating materials property predictions using machine learning. *Scientific Reports*, 3:2810, 09 2013. DOI: 10.1038/srep02810

[38] R. Ramprasad, R. Batra, G. Pilania, A. Mannodi-Kanakkithodi, and C. Kim. Machine learning in materials informatics: Recent applications and prospects. *NPJ Computational Materials*, 3:54, 2017. DOI: 10.1038/s41524-017-0056-5

[39] S. R. Kalidindi and M. De Graef. Materials data science: Current status and future outlook. *Annual Review of Materials Research*, 45(1):171, 2015. DOI: 10.1146/annurev-matsci-070214-020844

[40] H. C. Herbol, W. Hu, P. Frazier, P. Clancy, and M. Poloczek. Efficient search of compositional space for hybrid organic—inorganic perovskites via Bayesian optimization. *NPJ Computational Materials*, 4:51, 2018. DOI: 10.1038/s41524-018-0106-7

[41] M. Yamawaki, M. Ohnishi, S. Ju, and J. Shiomi. Multifunctional structural design of graphene thermoelectrics by Bayesian optimization. *Science Advances*, 4(6), 2018. DOI: 10.1126/sciadv.aar4192

[42] P. V. Balachandran, S. R. Broderick, and K. Rajan. Identifying the inorganic gene for high—temperature piezoelectric perovskites through statistical learning. *Proc. of the Royal Society A*, 467(2132):2271, 2011. DOI: 10.1098/rspa.2010.0543

[43] B. Meredig, A. Agrawal, S. Kirklin, J. E. Saal, J. W. Doak, A. Thompson, K. Zhang, A. Choudhary, and C. Wolverton. Combinatorial screening for new materials in unconstrained composition space with machine learning. *Physical Review B*, 89:094104, Mar 2014. DOI: 10.1103/physrevb.89.094104 68

[44] D. Xue, P. V. Balachandran, J. Hogden, J. Theiler, D. Xue, and T. Lookman. Accelerated search for materials with targeted properties by adaptive design. *Nature Communications*, 7:11241, April 2016. DOI: 10.1038/ncomms11241 68

[45] P. Raccuglia, K. C. Elbert, P. D. F. Adler, C. Falk, M. B. Wenny, A. Mollo, M. Zeller, S. A. Friedler, J. Schrier, and A. J. Norquist. Machine-learning-assisted materials discovery using failed experiments. *Nature*, 533(7601):73, May 2016. DOI: 10.1038/nature17439

[46] V. Duros, J. Grizou, W. Xuan, Z. Hosni, D.-L. Long, H. N. Miras, and L. Cronin. Human vs. robots in the discovery and crystallization of gigantic polyoxometalates. *Angewandte Chemie International Edition*, 56:10815, 2017. DOI: 10.1002/ange.201705721

[47] D. Xue, P. V. Balachandran, R. Yuan, T. Hu, X. Qian, E. R. Dougherty, and T. Lookman. Accelerated search for $BaTiO_3$-based piezoelectrics with vertical morphotropic phase boundary using Bayesian learning. *Proc. of the National Academy of Sciences of the USA*, 113(47):13301, 2016. DOI: 10.1073/pnas.1607412113

[48] F. Ren, L. Ward, T. Williams, K. J. Laws, C. Wolverton, J. Hattrick-Simpers, and A. Mehta. Accelerated discovery of metallic glasses through iteration of machine learning and high-throughput experiments. *Science Advances*, 4(4), 2018. DOI: 10.1126/sciadv.aaq1566

[49] J. Gao, Y. Liu, Y. Wang, X. Hu, W. Yan, X. Ke, L. Zhong, Y. He, and X. Ren. Designing high dielectric permittivity material in barium titanate. *Journal of Physical Chemistry C*, 121(24):13106, 2017. DOI: 10.1021/acs.jpcc.7b04636

[50] R. Yuan, Z. Liu, P. V. Balachandran, D. Xue, Y. Zhou, X. Ding, J. Sun, D. Xue, and T. Lookman. Accelerated discovery of large electrostrains in $BaTiO_3$-based piezoelectrics using active learning. *Advanced Materials*, 30:1702884, 2018. DOI: 10.1002/adma.201702884

[51] P. V. Balachandran, B. Kowalski, A. Sehirlioglu, and T. Lookman. Experimental search for high-temperature ferroelectric perovskites guided by two-step machine learning. *Nature Communications*, 9:1668, 2018. DOI: 10.1038/s41467-018-03821-9 68

[52] G. Hautier, C. C. Fischer, A. Jain, T. Mueller, and G. Ceder. Finding nature's missing ternary oxide compounds using machine learning and density functional theory. *Chemistry of Materials*, 22(12):3762, 2010. DOI: 10.1021/cm100795d 68

[53] C. C. Fischer, K. J. Tibbetts, D. Morgan, and G. Ceder. Predicting crystal structure by merging data mining with quantum mechanics. *Nature Materials*, 5(8):641, 2006. DOI: 10.1038/nmat1691 69

[54] P. V. Balachandran, D. Puggioni, and J. M. Rondinelli. Crystal-chemistry guidelines for noncentrosymmetric A_2BO_4 ruddlesden-popper oxides. *Inorganic Chemistry*, 53(1):336–348, 2014. DOI: 10.1021/ic402283c 69

[55] P. S. Halasyamani and K. R. Poeppelmeier. Noncentrosymmetric oxides. *Chemistry of Materials*, 10(10):2753–2769, 1998. DOI: 10.1021/cm980140w 69

[56] J. K. Krushke. *Doing Bayesian Data Analysis: A Tutorial with R, JAGS, and Stan*, Academic Press, New York, 2011. 69

[57] B. Efron. Bayes's Theorem in the 21st century. *Science*, 340(6137):1177, 2013. DOI: 10.1126/science.1236536 69

[58] W. Sun, S. T. Dacek, S. P. Ong, G. Hautier, A. Jain, W. D. Richards, A. C. Gamst, K. A. Persson, and G. Ceder. The thermodynamic scale of inorganic crystalline metastability. *Science Advances*, 2:e1600225, 2016. DOI: 10.1126/sciadv.1600225 72, 73

[59] F. Legrain, J. Carrete, A. van Roekeghem, G. K. H. Madsen, and N. Mingo. Materials screening for the discovery of new half-heuslers: Machine learning vs. ab initio methods. *Journal of Physical Chemistry*, 122:625, 2018. DOI: 10.1021/acs.jpcb.7b05296 73

[60] P. V. Balachandran, A. E. Emory, J. E. Gubernatis, T. Lookman, C. Wolverton, and A. Zunger. Predictions of new ABO_3 perovskite compounds by combining machine learning and density functional theory. *Physical Review Materials*, 2:043802, 2018. DOI: 10.1103/physrevmaterials.2.043802 73

[61] L. Breiman. Random forests. *Machine Learning*, 45:5, 2001. DOI: 10.1515/9783110941975 73

[62] J. H. Friedman. Greedy function approximation: A gradient boosting machine. *Annals of Statistics*, 29:1189, 2001. 73

[63] C. J. Bartel, S. L. Millican, A. M. Deml, J. R. Rumptz, W. Tumas, A. W. Weimer, S. Lany, V. Stevanović, C. B. Musgrave, and A. M. Holder. Physical descriptor for the Gibbs energy of inorganic crystalline solids and temperature-dependent materials chemistry. *Nature Communications*, 9(1):4168, 2018. DOI: 10.1038/s41467-018-06682-4 73

[64] T. Hastie, R. Tibshirani, and J. Friedman. *The Elements of Statistical Learning*. Springer, New York, 2008. DOI: 10.1007/978-0-387-84858-7 74, 75, 76, 78, 79

[65] P. Flach. *Machine Learning: The Art and Science of Algorithms that Make Sense of Data*. Cambridge University Press, New York, 2012. DOI: 10.1017/cbo9780511973000 74, 75

[66] Z. Ivezić, A. J. Connolly, J. T. VanderPlas, and A. Gray. *Statistics, Data Mining and Machine Learning in Astronomy*. Princeton University Press, Princeton, NJ, 2014. DOI: 10.23943/princeton/9780691151687.001.0001 74, 75, 76, 78

[67] J. R. Quinlan. Induction of decision trees. *Machine Learning*, 1:81, 1986. DOI: 10.1007/bf00116251 74

[68] J. Elith, J. R. Leathwick, and T. Hastie. A working guide to boosted regression trees. *Journal of Animal Ecology*, 77:802, 2008. DOI: 10.1111/j.1365-2656.2008.01390.x 74

[69] F. Pedregosa, G. Varoquaux, A. Gramfort, V. Michel, B. Thirion, O. Grisel, M. Blondel, P. Prettenhofer, R. Weiss, V. Dubourg, J. Vanderplas, A. Passos, D. Cournapeau, M. Brucher, M. Perrot, and E. Duchesnay. Scikit-learn: Machine learning in Python. *Journal of Machine Learning Research*, 12:2825, 2011. 79

[70] R Developmewnt Core Team. *R: A Language and Environment for Statistical Computing*. R Foundation for Statistical Computing, Vienna, Austria, 2013. 79

CHAPTER 4

Learning with Small Databases

In contrast to the combinatorial approaches using high throughput calculations that generate large data sets on ideal systems at $T = 0$ and $P = 0$ discussed in Chapter 3 and elsewhere in this book, real materials problems typically involve several components, are often solid solutions, and contain defects at $T \neq 0$ and $P \neq 0$. For crystals however, electronic structure calculations, say, based on density functional theory (DFT), often return reliable structural information and useful estimates of such physical quantities as dielectric and elastic constants and, in general, these calculations cannot return information on important functionalities such as whether the material is a candidate for memory applications, neuromorphic computing, etc. Most functionalities are established and quantified by experiment. As experiments can be quite time consuming and expensive, often data on only relatively few well-characterized samples (between ten and in the hundreds) are available. Hence, it is important to consider approaches that we can tailor and use for small data sets. The relevant question then becomes how we can effectively learn from existing, limited data to guide the next experiments in order to minimize the number of materials synthesized and measurements needed to be carried out to find a material with better targeted properties than what we may have in the training data [1]. Because of the emphasis on design, efficiency, and costs, it should come as little surprise that industry, as well as application areas such as drug design and cancer genomics, are very much at the forefront in developing iterative feedback methods, that is, adaptive learning methods, to reduce the number of experiments or calculations needed [2].

In this chapter, we review methods and applications that guide the optimal experiments to be performed sequentially to find materials with a targeted response [3]. This effort thus makes contact with aspects of experimental design, which is a well-developed area within statistics [4, 6, 7] and surrogate-based modeling [8], the latter using ideas based on the value of information and selection strategies for policies [9]. The aim here is to iteratively hone in on that part of the often large parameter space where a material with the desired property may be found, thereby reducing the number of experiments needed. This approach is therefore a departure from usual high-throughput screening, motivated by electronic structure calculations, in which the emphasis is on down selecting the best candidates for further studies. The methods we describe fall within the scope of what is known as Bayesian Global Optimization (BGO) [10]. Here, we describe how these methods are now starting to be used in materials science. However, before that we first explain the term Bayesian, a method developed and stated in terms of products and integrals of products of probabilities, and one of its simplest applications, the Gaussian process, which is the basis for BGO.

4.1 BACKGROUND

4.1.1 BAYES THEOREM

For two sets of events A and B, a fundamental result of probability theory is Bayes's Theorem, which states

$$P(A|B) = \frac{P(B|A)P(A)}{P(B)}, \tag{4.1}$$

where $P(A|B)$ and $P(B|A)$ are their conditional probabilities and $P(A)$ and $P(B)$ are their individual probabilities. For data analysis [11], we usually write this theorem as

$$P(\text{Model}|\text{Data}) = \frac{P(\text{Data}|\text{Model})P(\text{Model})}{P(\text{Data})}. \tag{4.2}$$

$P(\text{Model}|\text{Data})$ is called the posterior distribution. It represents the probability of the model after taking the data into account. Knowing it constitutes knowing the complete probabilistic solution to the problem. $P(\text{Data}|\text{Model})$ is the likelihood function. It is the probability of the data given the model. $P(\text{Model})$ is called the prior. It represents what we know about the model before the data is taken into account. Lastly, $P(\text{Data})$ is called the evidence. It is the normalization of the posterior, that is, $P(\text{Data}) = \int_{\text{Model}} P(\text{Model}|\text{Data})$. A typical Bayesian method often seeks adjusts the parameters of the Model so given the data the likelihood is maximized. If the likelihood is described by a Gaussian, this maximum is the mean (average) likelihood. The width about the maximum, the variance of the Gaussian, serves as a measure of the variance of the maximum. The maximum is taken as the optimal design of the model.

4.1.2 GAUSSIAN PROCESSES

Next, we describe a Gaussian process, which is in the class referred to as Bayesian methods. It is frequently encountered in physics and statistics, and we use it in the context of BGO. It is a collection of random variables such that any finite collection of them has a multi-variate Gaussian distribution. These methods do not target a best fit but instead compute a posterior (probability) distribution over models. The latter provides a quantification of the uncertainty in the predictions of the model. The significance of assuming Gaussian distributions for regression problems is that we can exactly perform many of the required algebraic operations and integrations by executing simple linear algebra operations on vectors and matrices (see Appendix B). Here, we just summarize a few results of lengthy analyses [12].

We assume our data, $\mathcal{D} = \{(\vec{x}_i, y_i)|i = 1, N\}$, is drawn from some underlying probability distribution $f(\vec{x})$ where $\vec{x}$ is a vector whose components represent the known values of the features chosen for the problem. A Gaussian noise model $\mathcal{N}(0, \sigma_n)$ represents the error η_i associated with the measurements or predictions of y_i, our targeted physical observable,

$$y_i = f(\vec{x}_i) + \eta_i. \tag{4.3}$$

Given the data, what we want is an estimate of the mean value y^* of y for a proposed set of features $\vec{x}^*$. In terms of probability theory, we seek the conditional probability $P(y^*|\vec{x}^*, \mathcal{D})$. For a Gaussian process, we can show that

$$P(y^*|\vec{x}^*, \mathcal{D}) = \mathcal{N}(\mu, \Sigma), \tag{4.4}$$

where the mean μ and covariance matrix Σ of this Gaussian distribution are

$$\begin{aligned} \mu &= K_{xx^*} K_{xx}^{-1} y \\ \Sigma &= K_{x^*x^*} - K_{xx}^T K_{xx^*}. \end{aligned}$$

The K's in turn are block matrices computed from a covariance matrix function

$$K = \begin{pmatrix} K_{xx} & K_{xx^*} \\ K_{xx^*}^T & K_{x^*x^*} \end{pmatrix} \tag{4.5}$$

with the matrix K_{xx} expressing the covariances among the observed values of the features, the matrix K_{xx^*} between the observations and the proposed feature, and the matrix, $K_{x^*x^*}$ between the proposed features. In essence,

$$\begin{bmatrix} \vec{y} \\ y^* \end{bmatrix} \sim \mathcal{N}(\vec{0}, K). \tag{4.6}$$

The covariance matrix K, usually called the kernel, is some assumed function that usually depends on the displacement $\vec{d} = \vec{x} - \vec{x}'$ in the space of features. For example, a simple, natural, but not always the most common choice is

$$K(\vec{d}) = \sigma^2 \exp\left(-\frac{|\vec{d}|^2}{2\ell^2}\right) + \sigma_n \delta(\vec{x} - \vec{x}'). \tag{4.7}$$

After the parameters of the kernel, in the above case σ, σ_n, and ℓ, are adjusted to the data by a procedure called the maximum likelihood method, we can insert the kernel and the proposed $\vec{x}^*$ into (4.4), find μ, our estimate for the new y^*, and use Σ as the estimate of its uncertainty.

4.2   APPLICATIONS

Before assessing the implications of limited data on materials applications, it is useful to recognize that in the field of bioinformatics the problem of small sample sizes and a large descriptor or feature space (in the hundreds and thousands if we consider genes) is well known [13]. For example, classifiers have to be trained in order to predict different types of breast cancers using aspects of gene signatures as the features, and there is a tradition in that field of using probabilistic Bayesian methods that incorporate prior knowledge of genes are turned on and off. It has been realized that such an approach has merit for developing classifiers and regressors and their associated errors, as opposed to merely using standard cross validation methods that are often not reliable particularly for small data sizes [13] unless the data are studied exceedingly thoroughly under stringent and robust conditions. By contrast, in materials informatics, it is only recently that researchers have undertaken in-depth analyses of regressor and classifier performance on materials data sets. The role and importance of uncertainties in predictions is also only just beginning to be appreciated [14–16].

The analysis that we will describe in this chapter goes beyond regression or classification studies, which are important in training inference models from data, but these do not address the problem of designing future experiments to identify a material with a given label or property. Figure 4.1 shows the adaptive design or active learning approach that contains aspects of optimal experimental design beyond merely training a "surrogate" or inference model, as in classification and regression. A dataset of features and properties serves as input to a surrogate model, which can be based on machine learning based methods and/or a reduced order or physics based model. The predictions and their uncertainties then flow to an experimental design component that selects the next experiment or calculation based on certain criteria. The new results then go into augmenting the training dataset and the loop repeats itself until the desired targets are met. The assumption is that by iterating this loop, we will converge to a solution better than in the initial training set. The problem can be stated as optimizing (say maximize for our purposes here) $f(x)$, that is, to find the design or material represented by the features x that maximizes f. The idea is to use some surrogate or inference model (off-the-shelf machine learning tools such as support vector or kernel ridge regression, decision trees or gradient tree boosting) to make the "best" prediction. The problem can be that of binary classification or optimizing a property, such as a Curie temperature, in which case to construct the surrogate we need a regression method, such as support vector regression (SVR) [17–19], to fit the features parameterizing the data to the property. If all we were doing is machine learning, we would then use the learned model to make a prediction. However, this prediction will not necessarily be optimal as it would only "exploit" the model's prediction. If we think of a cost function landscape for the property in the space of features, and if the cost function is not convex, such a "best value" prediction would likely correspond to a local minimum. To minimize the number of experiments, we need a means to "explore" this landscape beyond the confines of the space of the data by choosing a better next experiment than merely using the best prediction from the model [20]. In some sense, the surrogate alone allows us to interpolate between the available data points, but what we are striving for is a means to explore (extrapolate) beyond the points available or where we are most certain.

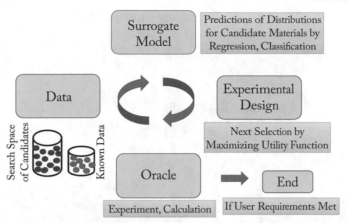

Figure 4.1: An active learning loop for finding optimal targets includes a surrogate model learned from data and an experimental design component with a utility function that encodes the criteria for selecting the next experiment or calculation. The space of possible, allowed compounds are ranked according to their expected utility and the compound with the maximum expected utility is chosen to test next. The new results augment the training data and the loop iterates sequentially.

The ideas discussed here are part of the field of experimental design, part of statistical design, which in its classical formulation studies how system's behave in which there are variations in the data that affect inputs and outputs [7]. The well-known factorial and composite designs in statistics typically serve as heuristics for exploring the relationship between the input factors and response variables. The idea is to obtain the interrelationships among all the different factors and their impact on the output to prioritize which set of experiments should be performed next. The concept of the value of the information (or cost of uncertainty) that is gained (or reduced) by observing a particular experiment or data point can serve as a guiding principle for undertaking the task. The possibilities of choosing certain experiments can then be ranked by the expected value of the information they provide, thereby helping to prioritize the experiments [3].

4.2.1   BAYESIAN GLOBAL OPTIMIZATION

Our previous description can be stated formally as a two-step process for the active learning loop of Figure 4.1 that involves (1) construction of the surrogate model and (2) definition of a utility function, previously introduced in decision theory and related to the value of information (and hence uncertainties) that facilitates optimal sampling decisions [21, 22]. This two-step approach is also known as BGO (Figure 4.2) if the surrogate model is replaced by a Gaussian process (see Appendix C). It has been found that a Bayesian approach for the surrogate model, where $f(x)$ is sampled from a prior distribution that assumes smoothness and locality, is quite a powerful

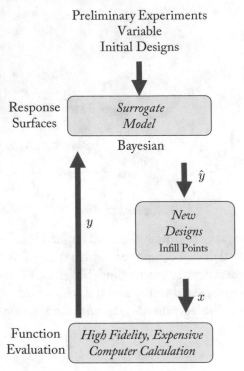

Figure 4.2: Bayesian Global Optimization uses a surrogate model based on data with input x from which new design points for the next calculations are selected with estimates for output y.

means to locate the extrema of unknown functions, especially when they are costly to evaluate and if information on their derivatives and convexity behavior is not available. Thus, assuming a Gaussian prior has been demonstrated to be quite efficient in guiding successive sampling [23]. The predicted mean and variance of the Gaussian process are then used as the input to the utility function, which prioritizes the basis of the decision process of what steps to take next in terms of performing a particular experiment or calculation [24]. The choice of next experiment is based on maximizing the expected utility function from the list of the many allowed possibilities. If the Gaussian process is not used as a surrogate, the mean and variance can be estimated from the data by the method of bootstrap sampling with replacement to mimic the properties of the population drawn from the original data sample [25]. Engineers often employ cheap surrogate models when the engineering model is computationally intense. The surrogate model is usually based on a relatively small number of training data points and Bayesian Global Optimization (BGO) has had quite an impact in industry in the study of large scale computational tasks, following the work of Jones, Schonlau, and Welch, who initially introduced the Efficient Global Optimization (EGO) method [1]. This work relied on early work by Kushner [26] on optimizing unknown functions, as well as on work by Mockus et al. [27]. The idea of maximizing the probability of improvement from the "best-so-far" was introduced in these pioneering studies. BGO has been used extensively in design applications when optimizing unknown functions, especially where the computational tasks are expensive [2].

If $\mathcal{D} = \{x_i, f(x_i)\}$ $(i = 1 : n)$ is the set of observations of the data, then by Bayes' rule, the posterior distribution is given by $P(f|\mathcal{D}) \propto P(\mathcal{D}|f)P(f)$, where the posterior will contain the updated beliefs about the unknown $f(x)$ given the prior information about f encoded in a distribution combined with how likely is the data seen so far, $P(\mathcal{D}|f)$, the likelihood function. This step is often interpreted as estimating $f(x)$ with a response surface or even a surrogate model, such as a Gaussian process (GP) or any machine learned model. A common example of this is a random walk with random step sizes and this stochastic process was used in the early work of Kushner [26]. However, a GP, the extension of the multivariate Gaussian distribution to a stochastic process with an infinite number of variables, has been shown to be well suited if we make a number of simplifying assumptions and has become standard for BGO [23, 28]. Instead of a Gaussian distribution over a random variable specified by a mean and variance, a GP is a distribution over functions specified by its mean function, m and covariance function k, that is, $f(x) \sim \mathcal{GP}(m(x), k(x, x')$. A common choice for k is $k(x, x') = \exp(-\frac{1}{2}||x - x'||^2)$, the squared exponential, which approaches 1 as data points x and x' are closer together and 0 if far apart. However, the GP can allow for several kernels. In addition to the Gaussian kernel $\exp(-1/2(h/\theta)^2)$ with $h = ||x - x'||$ and θ a parameter, the exponential $\exp(-h/\theta)$ and Matern functions $(1 + \sqrt{(3)}h/\theta) \exp(-\sqrt{(3)}h/\theta)$, $(1 + \sqrt{(5)}h/\theta + (1/3)5(h/\theta)^2) \exp(-\sqrt{(5)}h/\theta)$ and $\exp(-(h/\theta)^p)$ are also often used. In principle, the choice of these priors can affect the surrogate model predictions. However, this depends on the data sets.

4.2.2 EXAMPLES OF UTILITY FUNCTIONS

The second step of the active learning loop is the choice of the utility function and, while these ideas date back several decades to aspects of information and decision theory [29, 30], only recently have they been applied to problems related to materials discovery. Once we have chosen the utility function, $u(x|\mathcal{D}_{1:t-1})$, then its maximum over the all the data in the sampling domain, $x_t = \text{argmax}_x u(x|\mathcal{D}_{1:t-1})$, provides us with a potentially high value of the objective at which we can evaluate the function for the next experiment or calculation. This is the "good" decision making step to direct future sampling.

Expected Improvement (E[I])

The utility function we first describe is one of the best known and is an improvement-based function, I, initially suggested by Kushner [26], in order to maximize the probability of improvement over the best observed so far $f(x^+)$; that is, $I = \max(f(x) - f(x^+), 0)$ would be the improvement associated with each choice x. Before evaluating $f(x)$, we do not know what the improvement will be but as we have the probability distribution on $f(x)$, we can evaluate the amount of improvement using $E[I(x)] = E[f(x) - f(x^+)]$ for positive I, where E is the expectation over the posterior distribution [1]. As $f(x)$ is normally distributed with mean, μ, and standard deviation, σ, it can be shown that

$$
\begin{aligned}
E[I(x)] &= \int_{f(x^+)}^{\infty} (z - f(x^+))\phi(z)dz \\
&= (\mu(x) - f(x^+))\left[\phi\left(\frac{\mu(x) - f(x^+)}{\sigma(x)}\right) + \sigma(x)\Phi\left(\frac{\mu(x) - f(x^+)}{\sigma(x)}\right)\right].
\end{aligned}
\tag{4.8}
$$

Therefore, a new sample point, x^*, is chosen from other data points based on the largest expected improvement, that is, $x^* = \operatorname{argmax}_x EI[x]$. The limiting cases of uncertainty on $E[I(x)]$ give us the following.

- Small σ: $E[I(x)] \to \mu - f(x^+)$, i.e., choose the μ greater than $f(x^+)$ to maximize $E[I(x)]$ or *exploit* the model predictions.

- Large σ: $E[I(x)] \to \sigma$, that is, choose the μ with largest uncertainty to maximize $E[I(x)]$ or *explore* the sample space.

Moreover, $E[I(x)]$ is largest where both $\mu(x)$ and $\sigma(x)$ are large; that is, points likely to give the most gains would be those with the best prediction but also where uncertainty is greatest. These aspects emphasize the *exploration-exploitation* tradeoff, which is central to aspects of active [31] and reinforcement learning [32]. It characterizes many problems, including the multi-armed bandit problem, in which decisions need to be made repeatedly but where the outcomes are uncertain and we wish to obtain immediate results as to where to sample optimally now so that we get better results in the future. The multi-armed bandit problem [33] is a well-known example illustrating the exploration-exploitation trade-off and refers to a slot machine with n-arms or "bandits" in which each arm has its own probability distribution of success. By pulling any one of the arms, one gets a stochastic reward of either success or failure. The idea is to pull the arms in sequence one-by-one such that the total reward collected is maximized in the long run. This problem has applications in several real-world situations where we would like to select the "best" bandit out of a group of bandits, line-up optimization and evaluating social media influence.

Knowledge Gradient (KG)

A generalization of expected improvement in the presence of noise in the data is the knowledge gradient, $KG(x)$ [34, 35] utility function. With noise, the $f(x)$ values are now not known exactly. Instead, we consider $\mu(x)$ so that best choice in the following step with the largest improvement is given by $\mu_{n+1}^* = \max_x \mu_{n+1}(x)$ and $KG(x) = E[\mu_{n+1}^* - \mu_n^*]$. By normalizing the change in means by the number of standard deviations, σ, then $KG(x)$ is given by

$$
KG(x) = \sigma\left[\phi\left(\frac{\mu(x) - f(x^+)}{\sigma(x)}\right)\right] + \Phi\left(\frac{\mu(x) - f(x^+)}{\sigma(x)}\right).
\tag{4.9}
$$

Thus, the next sampling point is the one for which $KG(x)$ is largest, that is $\text{argmax}_x KG(x)$. The sampling region over which μ_n^*, μ_{n+1}^* are determined is not restricted to the observed data, as in the case of $E[I(x)]$, but can cover many allowed possibilities. Recently, expected improvement was generalized to also include noise [36].

Mean Objective Cost of Uncertainty (MOCU)

Recently, a utility function with an underlying objective-based uncertainty quantification scheme was introduced to study interactions among genes in gene regulatory networks [37–39] in order to develop drugs against cancers. The essential idea is that the presence of large uncertainties can negatively impact experimental design and the aim of new measurements should be to reduce uncertainties when searching for materials with desirable properties. Thus, one is striving for an objective-based uncertainty quantification scheme as compared to reducing overall variances or entropy as surrogates for information gain in an experiment. We assume a prior distribution that contains knowledge about the unknown parameters or features, say θ. This is updated to a posterior distribution after knowing the result of the experiment. If further measurements are required, then this distribution serves as the new prior for the next experimental design loop. The mean objective cost of uncertainty (MOCU) forms the utility function and measures the deterioration due to the presence of model uncertainty in the overall experimental design; that is, it measures the degradation in performance between an optimal design based on partial prior knowledge and data compared to a design with full knowledge of the system. Thus, the next experiment is chosen in such a way to shrink the uncertainty the most; that is, it is the experiment expected to minimize the variance in the posterior distribution the most.

Elaborating, suppose we are at iteration zero and we have a cost function $f(x)$ with unknown parameters θ; that is, we have observed initial data points and have updated prior distribution to posterior, $f(x)$, but not yet selected the next experiment x. If we want to stop doing any further measurements at this step and select the x that maximizes $f(x)$ based on our current state of knowledge or belief, we would select x which maximizes the expected value of $f(x)$ over the unknown parameters, θ, assuming we have a prior distribution for θ to evaluate the expected value. Thus, we would select $x_{\text{robust}} = \text{argmax}_x E_\theta f(x)$. This is the best "on average" we can do, hence the term "robust," because the θ's are not known and we are not guaranteed to choose the true optimal. It is this "robust" we seek which is expected to perform optimally *on average*. Now, corresponding to each parameter value, θ, the "optimal" material, x^+, can be found from $x^+ = \text{argmax}_x f_\theta(x)$. We will have an optimal for every parameter value θ and this is our best choice if unknown parameters do not exist. As we do not know the true parameter values θ, our loss is the difference between the true maximum value $f(x^+)$ and the value associated with the *robust* selection. This difference in the absence and presence of uncertainty is the Objective Cost of Uncertainty (OCU) and its expectation value over all θ is MOCU, the mean objective cost, which is given by

$$\text{MOCU} = E_\theta \left[f_\theta(x^+) - f_\theta(x_{\text{robust}}) \right]. \tag{4.10}$$

Thus, the next experiment selected is one which reduces MOCU the most. Hence, for each x, we consider all possible values of the experimental outcome y to calculate MOCU weighted by the probability distribution $P(y|x)$ of y, which is the true experiment outcome of x. The selected experiment for measurement, x^*, is then given by

$$x^* = \text{argmin}_x E_{y|x}\big(E_{\theta|x}[f_{\theta|x}(x^+) - f_{\theta|x}(x_{\text{robust}})]\big), \qquad (4.11)$$

where $E_{y|x}$ is expectation over $P(y|x)$. If we want to stop measurements at the next time step and select the candidate with highest expected $y = f(x)$, our final selection will be given by the *robust* alternative by taking the expectation over θ.

This method has recently been applied within phase field simulations to find dopants that minimize the energy dissipation, which affects the fatigue of shape memory alloys such as FePd [15]. These simulations typically assume an initial random dopant distribution for the FePd matrix and associated with the dopants is a concentration, potency and range, all of which are variable. The simulations are carried out such that a stress is applied in increments to obtain a stress vs. strain curve from which the energy dissipated is calculated. The data from the simulations is then used as the training data to fit a nonlinear model for the dissipation in terms of the variables: concentration, potency, and range of the dopants. This model is then applied to the experimental design step shown in Figure 4.3 to select the next best dopant and its features according to MOCU, that is, to search for the dopant that minimizes the total deterioration in the model uncertainty. After the measurement or experiment is performed (in the real situation) or as here, after the outcome is known, the prior distribution in terms of the unknown dopant parameters is then updated to obtain a posterior distribution, which becomes the new prior for the subsequent step in the active learning loop. The prior distribution is chosen as a Dirichlet distribution by generating different samples using different dopant concentrations, strength and parameters. The plot in Figure 4.4 shows how the average energy dissipation behaves as a function of numbers of measurements. The five measurements for the proposed approach are compared to ten measurements for pure exploitation and random selection selection policies. A summary of the utility functions discussed in this chapter is provided in Table 4.1.

4.2.3   APPLICATIONS TO MATERIALS SCIENCE

There are several examples in which the methodologies outlined above have been used to discover new materials as well as accelerate computational codes. We will discuss examples that the authors have worked on and then briefly discuss other work.

As discussed in detail in other chapters, one of the common methods in computational physics is to use DFT to calculate properties such as the band gap (E_g), thermal conductivity, and modulus (elastic, bulk, and shear) to name a few, in a high-throughput manner [40–43] by using available computational resources. The idea is to do a whole set of automated calculations for many compounds rather than on a case by case basis. More recently, there has been much interest in using machine learning methods to learn from DFT data. [41, 44–47]. Typically,

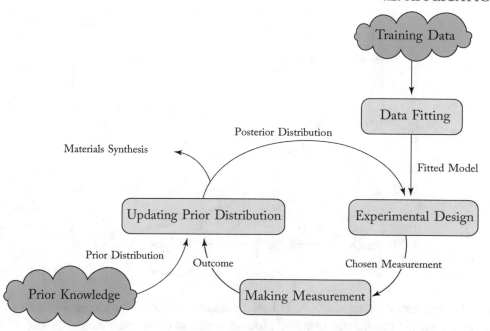

Figure 4.3: The Mean Objective Cost of Uncertainty (MOCU) is an objective-based uncertainty approach to experimental design in which a prior distribution serves to initiate the search. For the shape memory alloy application, simulations were initially performed to generate data from which a model was constructed that is the basis for the experimental design [15].

Table 4.1: Utility functions. A descriptive summary of the three main utility functions discussed in this review.

Utility Function	Description
Efficient global optimization	Evaluates trade-off between exploration and exploitation in the *absence* of uncertainties in the response
Knowledge gradient	Evaluates trade-off between exploration and exploitation in the *presence* of uncertainties in the response
Mean objective cost of uncertainty	Evaluates the degradation in performance between an optimal design based on partial prior knowledge and data compared to a design with full knowledge of the system

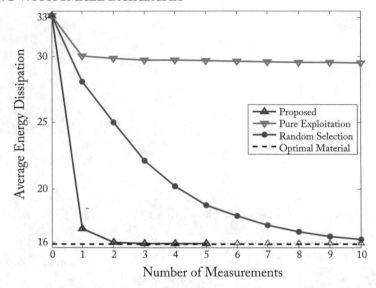

Figure 4.4: The performance of MOCU based on simulations of the shape memory alloy are shown compared to random selection and pure exploitation [15]. The y axis is the objective, the average energy dissipation for the alloy, and is plotted as a function of the number of sequential measurements. MOCU finds the optimal features characterizing the alloy in a few measurements.

machine learning combined with DFT involves generating a fairly large dataset by performing high-throughput calculations, training a machine-learned model on the dataset, and then using that model to predict new responses or results for compositions not present or represented in the dataset. This often does not account for uncertainties or active learning. Here, we demonstrate a specific application of active learning where machine learning methods iteratively guide DFT calculations toward promising regions in the composition space [48].

We will consider compounds known as apatites with the chemical formula $A_{10}(BO_4)_6X_2$, where A and B are divalent and pentavalent cations, respectively, and X is an anion. Apatites have applications as biomaterials, luminescent materials, and host lattices for immobilizing heavy and toxic elements and radiation tolerant materials [49]. In particular, we will study the family of compounds given by A={Mg, Ca, Sr, Ba, Zn, Cd, Hg or Pb}, B={P, As or V}, and X={F, Cl, Br or OH} with a total space of 96 unique compositions that satisfy the $A_{10}(BO_4)_6X_2$ stoichiometry. The objective then is to find the composition with the largest E_g in as few iterations or loops as possible by a combination of active learning methods and DFT calculations Figure 4.5a.

The training data set for this problem consisted of 13 randomly chosen apatites for which E_g was calculated within the generalized gradient approximation (GGA). Each apatite is represented with r_A, r_B, and r_X which represent the Shannon ionic radius of A-, B-, and X-site

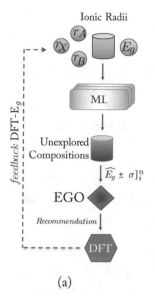

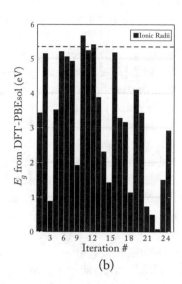

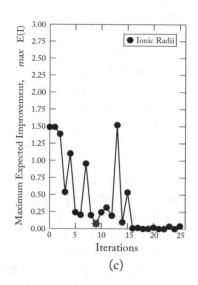

(a)   (b)   (c)

Figure 4.5: Example of how active learning can guide Density Functional Theory (DFT) calculations to minimize the number of calls that need to be made to the FDT code. (a) The overall strategy to guide the DFT calculations to find compositions giving rise to the largest band gap. (b) The band gap energies (E_g) obtained from DFT calculations (y-axis) plotted as a function of iteration number (x-axis). The red dashed line represents a composition with the largest E_g in the initial training data. (c) The behavior of the maximum E[I] obtained for each iteration of the loop is plotted as a function of iteration number. The data point corresponding to the best composition found in the 9th iteration need not coincide with the largest E[I] in the composition space.

elements, respectively [50]. The largest E_g (5.35 eV) in the training set was for the $Sr_{10}(PO_4)_6F_2$ (SrPF) composition in the $P6_3/m$ crystal symmetry. An ensemble of 100 SVR-RBF machine learning models were built to establish a relationship between the three features (r_A, r_B, and r_X) and the property, E_g. The hyperparameters for each SVR-RBF model were optimized using the leave-one-out cross-validation method. Each SVR-RBF model will return a prediction for E_g. Since there are 100 such models, the mean (μ or $\hat{E}_g$) and standard deviation (error bar, σ) can be estimated from the 100 E_g predictions.

The next step is to predict the $\hat{E}_g \pm \sigma$ for the remaining 83 compositions for which the E_g data from DFT-GGA calculation is not known, and here the EGO algorithm was used to recommend a composition that has the largest $E[I]$ for DFT-GGA validation and feedback. The training set is then augmented with this new composition. A computational "budget" of 25 total iterations is set to gain an understanding of the adaptive design process. In Figure 4.5b, the DFT-GGA calculated E_g data for the compositions recommended by EGO is shown. The optimal composition, $[Ca_{10}(PO_4)_6F_2]$, with the largest E_g of 5.67 eV was identified in the 9th iteration.

To provide insights into the iterative adaptive learning process, the max $E[I]$ at the end of each iteration was also tracked. The max $E[I]$ shows a non-monotonic trend and does not decrease smoothly as a function of number of iterations. This is shown in Figure 4.5c. The max $E[I]$ consistently attains a value of zero only from the 16th iteration onward. This is an intriguing insight because the optimal composition was found on the 9th iteration, yet the max $E[I]$ did not reduce to zero. This outcome sheds key light into the one of the difficult questions related to the stopping criterion, that is, when should one stop the iterative loop? We do not recommend stopping immediately after max $E[I]$ has reached a small value. Instead, it is important to confirm that the max $E[I]$ is consistently small and does not increase. This can be accomplished by running additional experiments or simulations a couple of additional times.

Recently, the discovery of a number of new alloys and ceramics has been accelerated using the feedback loop of Figure 4.1 via experimental synthesis and characterization instead of calculations. An example is the search for NiTi-based shape memory alloys with very small thermal hysteresis. Thermal hysteresis governs fatigue, and the idea is to find chemistries and compositions which will minimize thermal hysteresis. The approach of Xue et al. [51] assumes a family of alloys defined by $Ni_{50-x-y-z}Ti_{50}Cu_xFe_yPd_z$, where x, y, and z are compositions constrained by $50 - x - y - z \leq 30$, $x \leq 20$, $y \leq 5$, and $z \leq 20$ to avoid undesirable solid-solutions. The number of components here is less than five, but in principle more can be included at the expense of a larger search space of allowed possibilities. For training purposes, the authors assembled a database of their own measurements for compounds synthesized in an arc-melter under the same processing conditions and protocols. Each loop comprised four independent composition selections for synthesis and characterization by differential scanning calorimetry and x-ray diffraction. The models consisted of support vector regression and Gaussian processes, together with expected improvement that maximizes $E[I]$. They initiated the loop with 22 well-characterized training datasets from a palette of 800,000 allowed compositions with a composition control of 0.1% (Figure 4.6). The compound with the best performance was found to have a thermal hysteresis of 1.84 K, as measured by the peak-to-peak interval in the heat flow as a function of temperature on cooling and heating. By successively projecting the objective into a lower-dimensional feature space and carrying out DFT calculations, one of the conclusions of this work was that the composition of Fe in the new alloy was a key controlling factor leading to low hysteresis values. Of the 36 compounds synthesized and characterized, 14 had better performance than those in the initial training data. One can ask if these findings are the result of random occurrence. In statistics, a P-value (due to the eminent statistician R. A. Fisher [5]) is often used as a measure of statistical significance. For the alloy problem, it can be shown that $P < 0.001$, implying that the probability that the results are based purely on random chance is very small. The study compared the performance of several surrogate models and utility func-

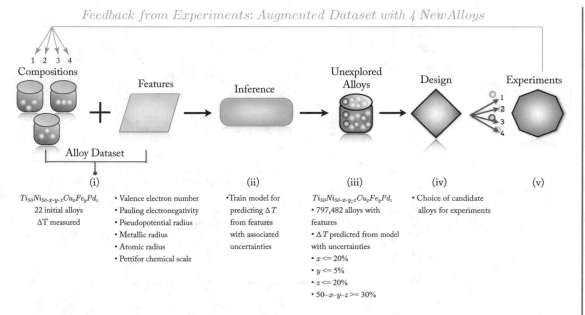

Figure 4.6: The adaptive design or active loop learning for finding shape memory alloys with the lowest thermal hysteresis. The training data including compositions and other materials features serve as input to building a surrogate model, the predictions and their uncertainties are then used by the experimental design component to select and rank candidates to test next. Four compositions were predicted in each iteration for synthesis and characterization [51].

tions and found the combination of an SVR for the surrogate model and the $E[I]$ for the utility function had the best performance on the training data.

As the transformation temperatures of alloys often determines the working window for device applications, Xue et al. [52] employed a polynomial model with features that capture chemical bonding and atomic size with design using EGO, KG, and maximum variance to predict the transformation temperature in shape memory alloys based on NiTi. The started with approximately 15 features and after a Pearson correlation analysis ended up with seven features on which they used a subset selection method to find the best combination of features with the lowest error using a linear model. Figure 4.7 shows that three features are quite adequate and the combination of weighted averages of en = Pauling electronegativity, ven = valence electron number, and dor = Waber-Cromer's pseudo potential radius of the constituents of the alloys fit the data best. In addition, as Figure 4.8 shows, an SVR model with a radial basis kernel (SVR.rbf) performs better than other modes investigated. However, a polynomial model for the transition temperature of the form $T_p = \beta_0 + \beta_1^{en} en + \beta_1^{dor} dor + \beta_1^{ven} ven + \beta_2^{dor} dor^2$, although not as good as SVR.rbf, provides valuable insight as it can be connected to a Landau-type model to understand the controlling factors, such as the lattice strains due to the dopants

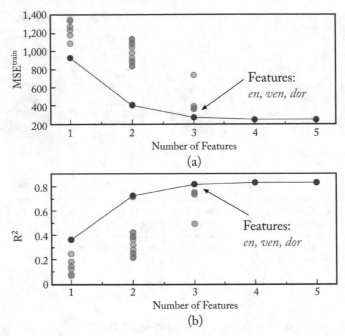

Figure 4.7: The mean squared error (MSEtrain), which captures the difference between the model predicted and measured values, and R^2 error between 0 and 1, for each possible linear model for subsets of the seven features in the training data for finding transition temperatures of alloys. The red frontier shows the best model for a given number of features, according to MSEtrain and R^2 and indicate little improvement beyond three features.

in the NiTi based SMA, that dictate transformation temperatures in NiTi-based alloys. The alloy $Ti_{50}Ni_{25}Pd_{25}$ was predicted to have the highest transition temperature of $189.56°C$, compared to the subsequent experimentally measured value of $182.89°C$. Recently, Ward et al. [53] considered a slightly different strategy in the search for ternary bulk metallic glasses by starting with data from the literature on glass forming ability related to amorphous ternary systems of the type Co-V-Zr. For this purpose, they used several models, including classification, whose predictions they essentially exploit in a staged approach to take into account for synthesis dependence. They conducted high-throughput experiments involving hundreds of data points based on parallel synthesis using magnetron co-sputtering.

One of the first applications of design to ferroelectrics was the use of Bayesian inference in conjunction with a Landau model for finding Pb-free $BaTiO_3$-based piezoelectric solid solutions with morphotropic phase boundaries (MPBs) that showed little temperature sensitivity [54] (Figure 4.9). The family of compounds considered belonging to the class of compounds $(Ba_{1-m}Ca_m)TiO_3$-$Ba(Zr_nTi_{1-n})O_3$ (BCT-BZT) and $(Ba_{1-m}Ca_m TiO_3$-

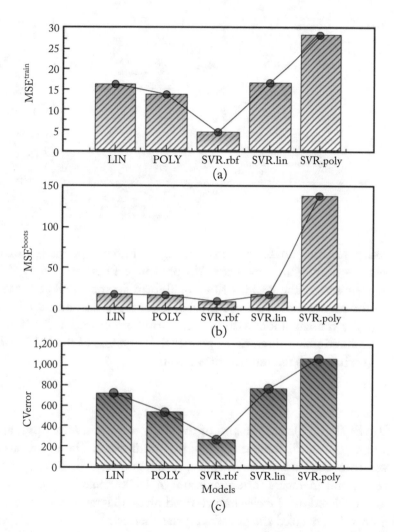

Figure 4.8: Performance of different models on the shape memory alloy training data in terms of MSEtrain, MSEboots, and CVerror. (a) MSEtrain, calculated using gives the training error for the model; (b) MSEboots (using 1000 bootstrap samples); and (c) CVerror (leave one out cross validation error) provide estimates of the test error.

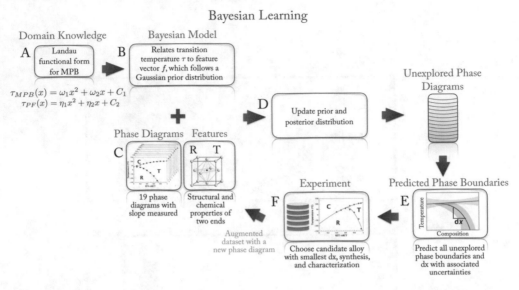

Figure 4.9: Bayesian learning for ferroelectrics using an initial experimental dataset of phase diagrams and features known to impact the Morphotropic Phase Boundary (MPB) between different ferrroelectric states. The Landau functional form for the MPB serves as prior knowledge to constrain the model and resulting Bayesian linear regression model relates the MPB to materials features. The trained model with updated prior and posterior distributions is applied to a dataset of unexplored phase diagrams to predict the curvature of the MPB. Minimizing the curvature, the target, increases the temperature sensitivity

$Ba(Sn_n Ti_{1-n})O_3$ (BCT-BTS), and the aim of the study was to find the optimal m and n such that the curvature of the MPB for the ceramic was minimized. The equations for the phase boundaries served to constrain the search space, which consisted of 1200 possible phase diagrams with different end members (or a maximum of 18,000 possible compositions), and the experimental data consisted of 19 well-characterized phase diagrams of synthesized solid solutions. Using features that coupled the atomic, crystal chemistry, and electronic structure properties of the two end members of the solid solutions, the authors used Bayesian regression, as well as a data-driven approach with EGO, to synthesize and characterize the solid solution $(Ba_{0.5}Ca_{0.5})TiO_3$-$Ba(Ti_{0.7}Zr_{0.3})O_3$. Figure 4.10 shows how the uncertainty is reduced in the updated Bayesian linear regression as compared to the initial predictions after an iteration. By studying correlations in the experimental data, Xue et al. [55] also show that the ratio of unit cell volumes and the ratio of ionic displacements of the tetragonal and rhombohedral ends influence the verticality of the MPB in both BaTiO-based and Pb-based systems and therefore can serve as excellent descriptors. As these quantities are not readily accessible for multi-dopant systems,

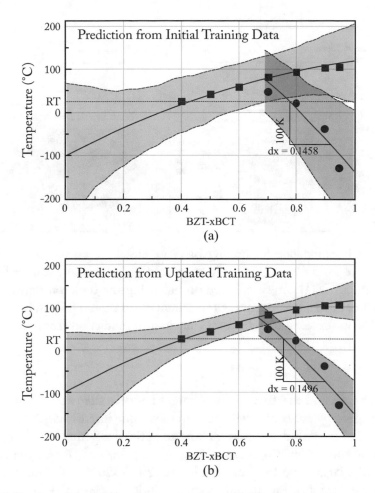

Figure 4.10: (A) Predicted (solid lines) vs. experimental (dots) phase diagram for the Pb-free BaTiO$_3$-based piezoelectric solid solution from Bayesian linear regression. The blue and red solid lines show the mean phase boundaries, and the blue and red dashed lines mark the 95% confidence intervals. (B) Updated Bayesian linear regression model after augmenting the experimental data. The uncertainties (dashed lines) are reduced using the updated model.

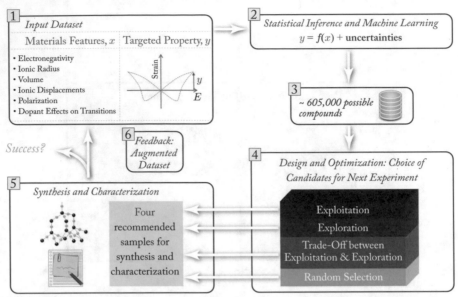

Figure 4.11: Active learning loop for finding $BaTiO_3$-based piezoelectrics with large strains at an applied voltage of 20 kV/cm. This study compared experimentally the performance of the expected improvement, E[I] criterion, based on balancing the trade-off between exploitation-exploration, to other design selection strategies, namely, maximum uncertainty from the surrogate model prediction (exploration), the best (maximum) prediction from the model (exploitation).

the ratios of ionic radii and effective nuclear charges between the two ends serve as surrogates for developing Pb-free piezoelectrics with high-temperature insensitivity.

Yuan et al. [56] recently demonstrated how an optimal experimental design approach can lead to the new piezoelectric $(Ba_{0.84}Ca_{0.16})(Ti_{0.90}Zr_{0.07}Sn_{0.03}O_3)$ with the largest electrostrain of 0.23% (unipolar strain 0.19%) under a field of 20 kV cm^{-1} in the barium titanate family (Figure 4.11). Their objective was to find a solid solution constrained to the family of compounds given by $Ba_{1-x-y}Ca_xSr_yTi_{1-u-v}Zr_uSn_vO_3$ with a large strain at an electric field of 20 kV/cm. Here, $x, y, u,$ and v are the mole fractions of specific dopants that obey $1 - x - y > 0.6$, $x < 0.4$, $y < 0.3$, $1 - u - v > 0.6$, $u < 0.3$, and $v < 0.3$. There are potentially about 605,000 possible compositions (controlled within 0.01%). Yuan et al.'s training data was 61 compounds that they synthesized under controlled conditions in their laboratory [56]. Clearly, this problem's large search space cannot be explored experimentally by just trial and error. The BGO design strategy with $E[I]$ found the optimal compound with a Sn composition of 3% in the third iteration and outperformed other strategies, including pure exploitation, exploration and random selection (Figure 4.12a). Figure 4.12b shows that the trend of the predicted strains are

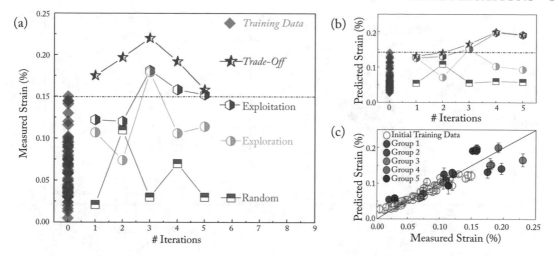

Figure 4.12: (a) A comparison of the results of experiments of the four selection strategies showing that the trade-off between exploration and exploitation performs better at each iteration than the other strategies in finding the compound with the largest strains. (b) The predicted behavior as a function of each iteration comparing the results from the four strategies. (c) Diagonal plot of measured vs. predicted showing the original training data and each subsequent group of four newly synthesized compounds in each iteration.

not too different than what is seen experimentally. The best compound is predicted to result on the fourth iteration instead of the third and the trade-off strategy outperforms the others. Also, the performance drops beyond the optimal, though not as significantly as seen experimentally. Figure 4.12c shows the original training data and all the groups of four compounds obtained in all six iterations on a diagonal plot, indicating that the model is not unreasonable. On either side of the 3 at% Sn composition, the strain decreases appreciably as seen in Figure 4.13b, suggesting that the Sn composition is at optimal. The improved performance of the 3% optimal compound is seen in the butterfly plot of strain vs. electric field (Figure 4.13a) when it is compared to the previous "best" compounds. By parametrizing a Landau free energy and performing phase-field simulations, they also show that the 3 at% Sn composition is associated with ease of switching of polarization domains in the tetragonal phase at room temperature relative to the compound without Sn. Thus, this work demonstrates that chemical substitutions in piezoelectrics can serve as an effective route to controlling the electrostrain properties of ferroelectric oxides at the domain level. A two-step strategy involving both classification and regression was also recently studied by Balachandran et al. [57] in the search for high Curie temperatures in Bi-based solutions that also include $Bi[M1_y M2_{1-y}]O_{3-(1-x)}PbTiO_3$ (Figure 4.14). The classification first separates one-phase stable compounds from other perovskites that are either unstable or contain multiple phases, and the regression and design guide experiments toward solutions with

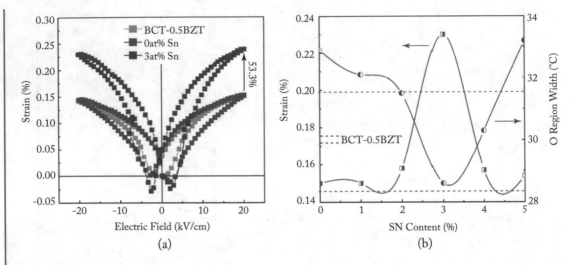

Figure 4.13: (a) The butterfly curve showing strain vs. electric field comparing the optimal 3 at% Sn compound with other well known compositions. (b) The values of measured strain (blue) as a function of Sn content, showing that the piezoelectric compound with 3 at% Sn compound is optimal. The red curve tracks the width of the region of stability of the orthorhombic phase, which decreases at 3 at% Sn composition.

high Curie temperatures. With this approach the authors found Bi-based perovskites containing novel dopant pairs, such as Fe,Co, NiSn, and CoAl at the B site, with high Curie temperatures (898 K for FeCo). These examples, and others in the literature, illustrate the efficacy of discovering new compounds with targeted properties in an active learning strategy involving experiments.

Several recent studies focused on learning from data generated using phenomenological codes. For example, Wang et al. [58] considered optimizing interphase properties in polymer nanocomposites by minimizing the difference between the predicted bulk property of a nanocomposite and that from the experimental data. For this purpose, they used a GP model as the surrogate in conjunction with $E[I]$. Their approach was tested on simulations of dielectric and viscoelastic properties in nanocomposites. In another example related to the design of light emitting diodes (LEDs), it is desirable to maximize electro-luminescence efficiency at high current densities. Leduc et al. [47, 59] considered an LED with five quantum wells where the indium compositions are the inputs (or features), and they optimized the quantum efficiency of the wells. They used a Gaussian process and $E[I]$ to show how to rapidly simulate the GaN-based LED structures in as few iterations as possible. The method rapidly converges to find a nearly optimal simulated LED efficiency in about 75 iterations, and subsequent iterations result in little improvement with attendant decreases in model uncertainty. Their design effectively

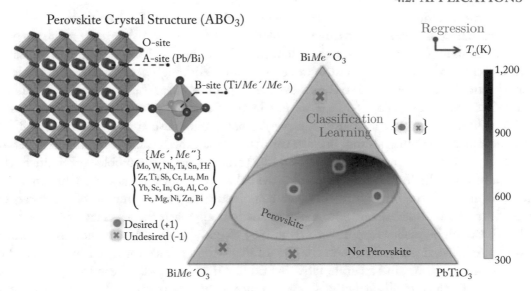

Figure 4.14: A two-step strategy for finding high-temperature ferroelectric perovskites in the Bi-based family of solid-solutions that include compounds belonging to the family $Bi[Me'_yMe''_{1-y}]O_{3-(1-x)}PbTiO_3$, where Me', Me'' are possible cations that can occupy the B site of the perovskite structure. The phase space consists of a total of 61,500 compositions and only about 0.28% are experimentally investigated. Classification learning methods are first used to isolate desired regions (chemistries and compositions) in the phase diagram where the candidate perovskite phases are expected to exist, and regression methods then predict the ferroelectric transition temperatures for the candidate perovskites.

guides the selection of the next LED structure to be examined based upon its expected efficiency as well as its model uncertainty. In particular, the active learning strategy produces a model that predicts the results of the Poisson-Schrödinger simulations of devices and simultaneously yields structures with higher simulated efficiencies.

As noted before, most studies have so far focused on optimizing single objectives whereas materials problems often have competing objectives. Thus, the ideas outlined here have also recently been applied to multi-objective optimization so that the next material identified in an optimization is chosen to improve on an existing Pareto front. This is discussed in greater detail in Chapter 5. In this regard, we note here that the generalization to multi-objectives of the surrogate-based optimization approach for single objectives was considered by Keane [60] and Svenson and Santner [61]. Recently, these algorithms were applied by Gopakumar et al. [62] applied these methods to several data sets already discussed, including SMAs, M_2AX phases and piezoelectrics, to assess the behavior of different selection strategies in the two-dimensional space of objectives. Also recently, a variation on this approach in terms of the hypervolume indicator was introduced for multiobjective optimization [63–65]. The hypervolume is a measure of the size of the space enclosed by all solutions on the Pareto front and a reference point given by the user. The expected improvement in hypervolume is then the gain and a measure of the closeness of a candidate point to the Pareto front. Although not as well studied, this approach is competitive compared to the usual expected improvement on the objectives as studied in materials applications [66, 67]. In terms of computational costs, typical implementations are considerably more resource intensive; however, recent studies have attempted to accelerate the computations of the integrals that arise in calculating the expected improvement in hypervolume [68]. As discussed in Chapter 6, related ideas are relevant to the problem of multi-fidelity optimization, in which the aim is to predict the behavior of the surrogate model from a large amount of inexpensive, but low-fidelity data combined with a small amount of expensive, but high-fidelity data. The idea here is to make predictions approach the accuracy of those associated with the high-fidelity data. Pilania et al. [69] recently applied this method to the case of band gaps evaluated using the much cheaper LDA method of electronic structure calculation and the more expensive, but accurate HSC hybrid functional method. The authors showed that the accuracy of the overall predictions depends on the relative fraction of the inexpensive vs. expensive data. Although the approaches used so far in materials problems are largely variants of co-Kriging, recently more sophisticated information fusion frameworks involving more than a pair of high-fidelity and low-fidelity models were used to design optimal microstructures [70].

Finally, we note that the No Free Lunch theorem provides a sobering constraint on the materials design problem. The theorem, discussed in Chapters 1 and 7, essentially states that there is no universal optimizer; that is, there is no guarantee that a model trained on a given data set will also apply to another data set. Thus, in the case of materials problems where data sizes can be limited, a purely data-driven approach has its limitations and will not, in general, be robust with well-defined uncertainties. In many of the studies we have reviewed, including those leading to new materials, the search spaces are generally relatively large and there are no clear stopping criteria. The design loop is terminated once a satisfactory outcome is achieved. Thus, moving forward, physics-based models will be essential in constraining the search space and providing adequate prior information for robust predictions as new data becomes available from experiments or calculations.

4.3   REFERENCES

[1] D. R. Jones, M. Schonlau, and W. J. Welch. Efficient global optimization of expensive black-box functions. *Journal of Global Optimization*, 13(4):455, 1998. DOI: 10.1007/s10898-004-0570-0 87, 92, 94

[2] A. I. J. Forrester, A. Sóbester, and A. J. Keane. *Engineering Design via Surrogate Modelling: A Practical Guide*. John Wiley, New York, 2008. DOI: 10.2514/4.479557 87, 93

[3] W. B. Powell and I. O. Ryzhov. *Optimal Learning*. Wiley, New York, 2012. DOI: 10.1002/9781118309858 87, 91

[4] R. A. Fisher. *The Design of Experiments*, 5th ed., Oliver and Boyd, Oxford, 1949. DOI: 10.2307/2277749 87

[5] R. A. Fisher. Statistical methods and scientific induction. *J.R. Stat. Soc. Ser. B. Stat. Methodol.*, 17, 69–78, 1955. 100

[6] G. E. P. Box and K. B. Wilson. On the experimental attainment of optimum conditions. *Journal of the Royal and Statistical Society B*, 13(1):1, 1951. DOI: 10.1007/978-1-4612-4380-9_23 87

[7] M. Cavazzuti. *Optimization Methods*. Springer-Verlag, Heidelberg, 2013. DOI: 10.1007/978-3-642-31187-1 87, 91

[8] A. I. J. Forrester, A. Sóbester, and A. J. Keane. *Engineering Design via Surrogate Modelling: A Practical Guide*. John Wiley, New York, 2008. DOI: 10.2514/4.479557 87

[9] D. V. Lindley. On a measure of the information provided by an experiment. *Annals of Mathematical Statistics*, 27(4):986, 1956. DOI: 10.1214/aoms/1177728069 87

[10] E. Brochu, V. M. Cora, and N. de Freitas. A tutorial on Bayesian optimization of expensive cost functions, with application to active user modeling and hierarchical reinforcement learning. *ArXiv 1012.2599*, 2010. 87

[11] D. S. Sivia and J. Skilling. *Data Analysis: A Bayesian Tutorial*. Oxford University Press, Oxford, 2006. 88

[12] C. E. Rasmussen and K. J. Williams. *Gaussian Processes for Machine Learning*. MIT Press, Cambridge, MA, 2006. DOI: 10.7551/mitpress/3206.001.0001 88

[13] L. A. Dalton and E. R. Dougherty. Optimal classifiers with minimum expected error within a bayesian framework—Part II: Properties and performance analysis. *Pattern Recognition*, 46(5):1288, 2013. DOI: 10.1016/j.patcog.2012.10.019 90

[14] X. Du and W. Chen. Efficient uncertainty analysis methods for multidisciplinary robust design. *AIAA Journal*, 40(3):545, 2002. DOI: 10.2514/3.15095 90

[15] R. Dehghannasiri, D. Xue, P. V. Balachandran, M. R. Yousefi, L. A. Dalton, T. Lookman, and E. R. Dougherty. Optimal experimental design for materials discovery. *Computational Materials Science*, 129:311, 2017. DOI: 10.1016/j.commatsci.2016.11.041 96, 97, 98

[16] J. Ling, M. Hutchinson, E. Antono, S. Paradiso, and B. Meredig. High-dimensional materials and process optimization using data-driven experimental design with well-calibrated uncertainty estimates. *Integrating Materials and Manufacturing Innovation*, 6(3):207, 2017. DOI: 10.1007/s40192-017-0098-z 90

[17] T. Hastie, R. Tibshirani, and J. Friedman. *The Elements of Statistical Learning*. Springer, New York, 2008. DOI: 10.1007/978-0-387-84858-7 90

[18] P. Flach. *Machine Learning: The Art and Science of Algorithms that Make Sense of Data*. Cambridge University Press, New York, 2012. DOI: 10.1017/cbo9780511973000

[19] Z. Ivezić, A. J. Connolly, J. T. VanderPlas, and A. Gray. *Statistics, Data Mining and Machine Learning in Astronomy*. Princeton University Press, Princeton, NJ, 2014. DOI: 10.23943/princeton/9780691151687.001.0001 90

[20] P. V. Balachandran, D. Xue, J. Theiler, J. Hogden, and T. Lookman. Adaptive strategies for materials design using uncertainties. *Scientific Reports*, 6:19660, 2016. DOI: 10.1038/srep19660 90

[21] M. A. Clyde. Experimental design: Bayesian designs, Eds., Neil J. Smelser and Paul B. Baltes, *International Encyclopedia of the Social and Behavioral Sciences*, pp. 5075–5081, Pergamon, 2001. DOI: 10.1016/B0-08-043076-7/00421-6 91

[22] E. Brochu, V. M. Cora, and N. de Freitas. A tutorial on Bayesian optimization of expensive cost functions, with application to active user modeling and hierarchical reinforcement learning. *ArXiv e-prints*, 2010. 91

[23] J. Mockus. Application of Bayesian approach to numerical methods of global and stochastic optimization. *Journal of Global Optimization*, 4(4):347, 1994. DOI: 10.1007/bf01099263 92, 93

[24] D. V. Lindley. On a measure of the information provided by an experiment. *Annals of Mathematical Statistics*, 27(4):986–1005, 1956. DOI: 10.1214/aoms/1177728069 92

[25] B. Efron and G. Gong. A leisurely look at the bootstrap, the jackknife, and cross-validation. *American Statistician*, 37(1):36, 1983. DOI: 10.2307/2685844 92

[26] H. J. Kushner. A new method of locating the maximum of an arbitrary multi-peak curve in the presence of noise. *Journal of Basic Engineering*, 86:97, 1964. DOI: 10.1115/1.3653121 92, 93

[27] J. Mockus, V. Tiesis, and A. Zilinskas. The application of Bayesian methods for seeking the extremum. In L. C. W. Dixon and G. P. Szego, Eds., *Towards Global Optimisation*, volume 2, p. 117, Elsevier, Amsterdam, 1978. 93

[28] A. O'Hagan and J. F. C. Kingman. Curve fitting and optimal design for prediction. *Journal of the Royal Statistical Society B*, 40(1):1, 1978. DOI: 10.1111/j.2517-6161.1978.tb01643.x 93

[29] D. V. Lindley and A. F. M. Smith. Bayes estimates for the linear model. *Journal of the Royal Statistical Society B*, 34(1):1, 1972. DOI: 10.1111/j.2517-6161.1972.tb00885.x 93

[30] R. A. Howard. Information value theory. *IEEE Transactions on Systems, Science, and Cybernetics*, 2:22, 1966. DOI: 10.1109/tssc.1966.300074 93

[31] L. P. Kaelbling, M. L. Littman, and A. W. Moore. Reinforcement learning: A survey. *J. Artificial Intelligence Research*, 4:237, 1996. DOI: 10.1613/jair.301 94

[32] R. S. Sutton. Learning to predict by the methods of temporal differences. *Machine Learning*, 3(1):9, 1988. DOI: 10.1007/bf00115009 94

[33] A. Slivkins. Introduction to multi-armed bandits. *Foundations and Trends in Machine Learning*, 12, 1–286, 2019. DOI: 10.1561/2200000068 94

[34] P. I. Frazier, W. B. Powell, and S. Dayanik. The knowledge gradient policy for correlated normal beliefs. *INFORMS Journal on Computing*, 21:599, 2009. DOI: 10.1287/ijoc.1080.0314 94

[35] W. Scott, P. I. Frazier, and W. B. Powell. The correlated knowledge gradient for simulation optimization of continuous parameters using Gaussian process regression. *SIAM Journal on Optimization*, 21:996, 2011. DOI: 10.1137/100801275 94

[36] B. Letham, B. Karrery, G. Ottoniz, and E. Bakshyx. Constrained Bayesian optimization with noisy experiments. *Bayesian Analysis*, 14:495, 2019. DOI: 10.1214/18-ba1110 95

[37] B. J. Yoon, X. Qian, and E. R. Dougherty. Quantifying the objective cost of uncertainty in complex dynamical systems. *IEEE Transactions on Signal Processing*, 61(9):2256, 2013. DOI: 10.1109/tsp.2013.2251336 95

[38] R. Dehghannasiri, B.-J. Yoon, and E. R. Dougherty. Optimal experimental design for gene regulatory networks in the presence of uncertainty. *IEEE/ACM Transactions on Computational Biology and Bioinformatics*, 12(4):938, 2015. DOI: 10.1109/tcbb.2014.2377733

[39] R. Dehghannasiri, B.-J. Yoon, and E. R. Dougherty. Efficient experimental design for uncertainty reduction in gene regulatory networks. *BMC Bioinformatics*, 16(13):S2, 2015. DOI: 10.1186/s12859-015-0839-y 95

[40] J. E. Saal, S. Kirklin, M. Aykol, B. Meredig, and C. Wolverton. Materials design and discovery with high-throughput density functional theory: The open quantum materials database (OQMD). *JOM*, 65(11):1501, 2013. DOI: 10.1007/s11837-013-0755-4 96

[41] V. Sharma, C. Wang, R. G. Lorenzini, R. Ma, Q. Zhu, D. W. Sinkovits, G. Pilania, A. R. Oganov, S. Kumar, G. A. Sotzing, S. A. Boggs, and R. Ramprasad. Rational design of all organic polymer dielectrics. *Nature Communications*, 5:4845, 2014. DOI: 10.1038/ncomms5845 96

[42] A. Jain, S. P. Ong, G. Hautier, W. Chen, W. D. Richards, S. Dacek, S. Cholia, D. Gunter, D. Skinner, G. Ceder, and K. A. Persson. Commentary: The materials project: A materials genome approach to accelerating materials innovation. *APL Materials*, 1(1), 2013. DOI: 10.1063/1.4812323

[43] S. Curtarolo, W. Setyawan, S. Wang, J. Xue, K. Yang, R. H. Taylor, L. J. Nelson, G. L. Hart, S. Sanvito, M. Buongiorno-Nardelli, N. Mingo, and O. Levy. AFLOWLIB.ORG: A distributed materials property repository from high-throughput *ab initio* calculations. *Computational Materials Science*, 58(1):227, 2012. DOI: 10.1016/j.commatsci.2012.02.002 96

[44] M. de Jong, W. Chen, R. Notestine, K. Persson, G. Ceder, A. Jain, M. Asta, and A. Gamst. A statistical learning framework for materials science: Application to elastic moduli of k-nary inorganic polycrystalline compounds. *Scientific Reports*, 6:34256, 10 2016. DOI: 10.1038/srep34256 96

[45] L. Ward, A. Agrawal, A. Choudhary, and C. Wolverton. A general-purpose machine learning framework for predicting properties of inorganic materials. *NPJ Computational Materials*, 2:16028, 08 2016. DOI: 10.1038/npjcompumats.2016.28

[46] A. Seko, H. Hayashi, K. Nakayama, A. Takahashi, and I. Tanaka. Representation of compounds for machine-learning prediction of physical properties. *Physical Review B*, 95:144110, 2017. DOI: 10.1103/physrevb.95.144110

[47] P. V. Balachandran, T. Shearman, J. Theiler, and T. Lookman. Predicting displacements of octahedral cations in ferroelectric perovskites using machine learning. *Acta Crystallographica B*, 73(5):962, 2017. DOI: 10.1107/s2052520617011945 96, 108

[48] P. V. Balachandran, D. Xue, J. Theiler, J. Hogden, J. E. Gubernatis, and T. Lookman. Importance of feature selection in machine learning and adaptive design for materials. In: Lookman T., Eidenbenz S., Alexander F., and Barnes C., Eds., *Materials Discovery and Design*, Springer Series in Materials Science, vol. 280, Springer, Cham, 2018. 98 DOI: 10.1007/978-3-319-99465-9_3

[49] T. White, C. Ferraris, J. Kim, and S. Madhavi. Apatite—An adaptive framework structure. *Reviews in Mineralogy and Geochemistry*, 57(1):307, 2005. DOI: 10.1515/9781501509513-010 98

[50] R. D. Shannon. Revised effective ionic radii and systematic studies of interatomic distances in halides and chalcogenides. *Acta Crystallographica A*, 32:751–767, 1976. DOI: 10.1107/s0567739476001551 99

[51] D. Xue, P. V. Balachandran, J. Hogden, J. Theiler, D. Xue, and T. Lookman. Accelerated search for materials with targeted properties by adaptive design. *Nature Communications*, 7:11241, 04 2016. DOI: 10.1038/ncomms11241 100, 101

[52] D. Xue, D. Xue, R. Yuan, Y. Zhou, P. V. Balachandran, X. Ding, J. Sun, and T. Lookman. An informatics approach to transformation temperatures of NiTi—based shape memory alloys. *Acta Materialia*, 125:532, 2017. DOI: 10.1016/j.actamat.2016.12.009 101

[53] F. Ren, L. Ward, T. Williams, K. J. Laws, C. Wolverton, J. Hattrick-Simpers, and A. Mehta. Accelerated discovery of metallic glasses through iteration of machine learning and high-throughput experiments. *Science Advances*, 4(4), 2018. DOI: 10.1126/sciadv.aaq1566 102

[54] D. Xue, P. V. Balachandran, R. Yuan, T. Hu, X. Qian, E. R. Dougherty, and T. Lookman. Accelerated search for $BaTiO_3$-based piezoelectrics with vertical morphotropic phase boundary using Bayesian learning. *Proc. of the National Academy of Sciences of the USA*, 113(47):13301, 2016. DOI: 10.1073/pnas.1607412113 102

[55] D. Xue, P. V. Balachandran, H. Wu, R. Yuan, Y. Zhou, X. Ding, J. Sun, and T. Lookman. Material descriptors for morphotropic phase boundary curvature in lead-free piezoelectrics. *Applied Physics Letters*, 111(3):032907, 2017. DOI: 10.1063/1.4990955 104

[56] R. Yuan, Z. Liu, P. V. Balachandran, D. Xue, Y. Zhou, X. Ding, J. Sun, D. Xue, and T. Lookman. Accelerated discovery of large electrostrains in $BaTiO_3$-based piezoelectrics using active learning. *Advanced Materials*, p. 1702884, 2018. DOI: 10.1002/adma.201702884 106

[57] P. V. Balachandran, B. Kowalski, A. Sehirlioglu, and T. Lookman. Experimental search for high-temperature ferroelectric perovskites guided by two-step machine learning. *Nature Communications*, 9(1):1668, 2018. DOI: 10.1038/s41467-018-03821-9 107

[58] Y. Wang, Y. Zhang, H. Zhao, X. Li, Y. Huang, L. S. Schadler, W. Chen, and L. C. Brinson. Identifying interphase properties in polymer nanocomposites using adaptive optimization. *Composites Science and Technology*, 162:146, 2018. DOI: 10.1016/j.compscitech.2018.04.017 108

[59] B. Rouet-Leduc, K. Barros, T. Lookman, and C. J. Humphreys. Optimisation of GaN LEDs and the reduction of efficiency droop using active machine learning. *Scientific Reports*, 6:24862, 2016. DOI: 10.1038/srep24862 108

[60] A. J. Keane. Statistical improvement criteria for use in multiobjective design optimization. *AIAA Journal*, 44(4):879, 2018/03/23 2006. DOI: 10.2514/1.16875 109

[61] J. Svenson and T. Santner. Multiobjective optimization of expensive-to-evaluate deterministic computer simulator models. *Computational Statistics and Data Analysis*, 94:250, 2016. DOI: 10.1016/j.csda.2015.08.011 109

[62] A. M. Gopakumar, P. V. Balachandran, D. Xue, J. E. Gubernatis, and T. Lookman. Multi-objective optimization for materials discovery via adaptive design. *Scientific Reports*, 8(1):3738, 2018. DOI: 10.1038/s41598-018-21936-3 109

[63] M. Emmerich, N. Beume, and B. Naujoks. An EMO algorithm using the hypervolume measure as selection criterion. In C. A. Coello, A. H. Aguirre, and E. Zitzler, Eds., *Evolutionary Multi-Criterion Optimization*, p. 62, Springer, Heidelberg, 2005. DOI: 10.1007/978-3-540-31880-4_5 110

[64] L. Lu and C. M. Anderson-Cook. Adapting the hypervolume quality indicator to quantify trade-offs and search efficiency for multiple criteria decision making using Pareto fronts. *Quality and Reliability Engineering International*, 29(8):1117, 2013. DOI: 10.1002/qre.1464

[65] Y. Cao, B. J. Smucker, and T. J. Robinson. On using the hypervolume indicator to compare Pareto fronts: Applications to multi-criteria optimal experimental design. *Journal of Statistical Planning and Inference*, 160:60, 2015. DOI: 10.1016/j.jspi.2014.12.004 110

[66] A. Talapatra, S. Boluki, T. Duong, X. Qian, E. Dougherty, and R. Arróyave. Autonomous efficient experiment design for materials discovery with Bayesian model averaging. *Physical Review Materials*, 2:113803, 2018. DOI: 10.1103/physrevmaterials.2.113803 110

[67] A. Solomou, G. Zhao, S. Boluki, J. K. Joy, X. Qian, I. Karaman, R. Arryave, and D. C. Lagoudas. Multi-objective Bayesian materials discovery: Application on the discovery of precipitation strengthened NiTi shape memory alloys through micromechanical modeling. *Materials and Design*, 160:810, 2018. DOI: 10.1016/j.matdes.2018.10.014 110

[68] G. Zhao, R. Arryave, and X. Qian. Fast exact computation of expected hypervolume improvement. unpublished, 2018. DOI: 10.1007/s10898-019-00798-7 110

[69] G. Pilania, J. E. Gubernatis, and T. Lookman. Multi-fidelity machine learning models for accurate bandgap predictions of solids. *Computational Materials Science*, 129:156, 2017. DOI: 10.1016/j.commatsci.2016.12.004 110

[70] S. F. Ghoreishi, S. Friedman, and D. L. Allaire. Adaptive dimensionality reduction for fast sequential optimization with Gaussian processes. *Journal of Mechanical Design*, March 2019. DOI: 10.1115/1.4043202 110

CHAPTER 5

Multi-Objective Learning

Real engineering problems are generally multi-objective where one or more properties must be simultaneously optimized. In a multi-objective optimization problem, we rarely have a unique solution that satisfies all the constraints associated with all the objectives. Instead, we settle for "trade-offs" and find acceptable solutions that balance the trade-off between the target objectives. Optimal materials design and discovery hinges on identifying and harnessing selected regions of a multidimensional property space that represents the best trade-offs among the different properties for a target application. In this chapter, we introduce key concepts in multi-objective optimization, such as Pareto dominance, optimality, and fronts, and discuss two basic approaches to solving such problems based on froward and inverse modeling. The former exploits the data; the latter is based on the exploitation-exploration approach discussed in Chapter 4 for materials design and discovery targeting a single objective.

5.1   INTRODUCTION

In the previous chapters, we explored machine learning (ML) methods and machine learning-based approaches with a focus on a single property of interest, that is, single objective optimization problems. However, real engineering problems are multi-objective in nature where one or more properties must be *simultaneously* optimized. In some cases, the two (or more) properties of interest can be related (directly or inversely), whereas in other cases the properties can be independent of one another. For example, many structural applications demand materials that have both high strength and high ductility. However, the two properties are inversely correlated to one another. Thus, metallurgists are faced with a challenge to develop novel innovative strategies that will result in a microstructure that enables one to maximize both strength and ductility in a structural material [1]. We already illustrate Ashby's plots in Chapter 1. Another such example is presented by dielectric materials for electrical energy storage. While the total electrical energy stored in a dielectric capacitor depends on both its dielectric constant and the ability of the insulating material to withstand high electric fields without failing to a metallic state (which in turn is governed by the materials electronic bandgap), the two properties are again well known to be inversely correlated. Therefore, our ability to design novel and improved electric energy storage materials largely depends on a combined optimization of the dielectric constant and the bandgap over a target class of materials [2, 3]. Furthermore, practical examples would also consider other relevant aspects dictating performance of a material for a given application such as processability, cost, and long-term stability.

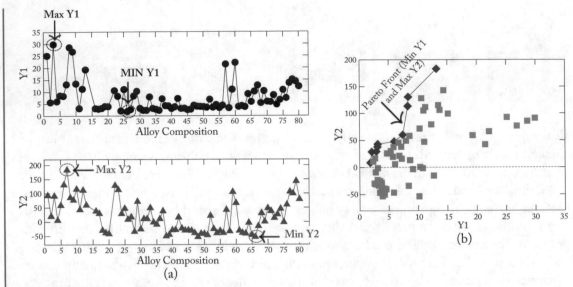

Figure 5.1: An illustration of a design challenge from the context of a bi-objective optimization problem. Target properties of interest are denoted as Y1 and Y2. (a) A plot between Y1 (Y2) as a function of alloy compositions. Note that the maxima/minima in Y1 and Y2 do not occur for the same composition. (b) A plot between Y1 and Y2 is shown and data points that satisfy the constraints of minimum Y1 and maximum Y2 are highlighted as red diamonds. The envelope that encompass these data points is called the "Pareto Front," which can be defined as a set of non-dominated solutions, being chosen as optimal, if no objective can be improved without sacrificing at least one other objective. Terms and definitions used are explained in Section 5.2.

It is important to recognize that in a multi-objective optimization scenario we rarely have one unique solution that will satisfy all constraints. Instead, we settle for "trade-offs" and find acceptable solutions that balance the trade-off between the target objectives. This is illustrated in Figure 5.1 using the data taken from the published work of Gopakumar et al. [4]. Figure 5.1a shows a plot of Y1 and Y2 as a function of alloy compositions, where Y1 and Y2 are the two target properties of interest. It can be seen that in both Y1 and Y2, the maximum and minimum do not coincide. In other words, we do not have a unique alloy composition where the maxima and minima in Y1 and Y2 occur together. As a result, only a trade-off between Y1 and Y2 will yield an "optimal" alloy composition in this dataset. One such scenario is depicted in Figure 5.1b, where the "hypothetical" bi-objective goal is to find alloy compositions that minimize Y1 and maximize Y2. The data points shown as red diamonds satisfy the bi-objective design requirements and represent the optimal solution set.

In fact, beyond the aforementioned two pedagogical cases, one can find numerous examples where optimal materials design hinges on identifying and harnessing selected regions of a multidimensional property space that represents the best trade-offs among the different properties for a target application. Thus, in many cases a research goal in materials science and engineering problems can be translated into a *multi-objective optimization* (MOO) problem [5–10]. The focus of this chapter is to provide an overview of some of the key concepts and state-of-the-art techniques (rooted in the principles of machine learning and optimization theory) that have been demonstrated in materials science applications. The intention is not to provide a comprehensive review, but highlight key concepts and methods that are most relevant for MOO from the context of applications in the domain of materials science and engineering.

5.2 DEFINITIONS AND KEY CONCEPTS

Formally, MOO can be defined as an optimization over a set of objective functions $\{f_i : \mathbb{R}^n \to \mathbb{R}, i = 1, ..., k\}$ as follows (here assuming a maximization problem for the sake of illustration):

$$\max \mathbf{f}(\mathbf{x}) := [f_1(\mathbf{x}), \ f_2(\mathbf{x}), \ldots, \ f_k(\mathbf{x})], \tag{5.1}$$

where $\mathbf{x} = [x_1, x_2, x_3, \ldots, x_n]^T$ represents a vector or n decision variables. Note that here $\mathbf{x}$ should not be confused with the feature vector employed in building the machine learning-based surrogate model in the first place. Based on physical arguments, or resorting to the underlying physics of the problem, often it becomes possible to define upper (x_i^u) and lower (x_i^l) bounds on each of the decision vector component, which then define the relevant search space $\mathcal{S}$. In practical applications, in addition to the objective functions one may also encounter some additional constraints $g_r(\mathbf{x}) \leq 0$ with $r = 1, 2, \ldots, m$ and $h_s(\mathbf{x}) = 0$ with with $s = 1, 2, \ldots, p$ which need to be obeyed. This further narrows the feasible region of the search space $\mathcal{S}$ to $\mathcal{F}$ such that $\mathcal{F} \subseteq \mathcal{S}$. Any vector of decision variables $\mathbf{x}$ which satisfies all the constraints is considered a feasible solution and therefore resides in $\mathcal{F}$.

The above description allows us to define important key concepts such as Pareto dominance, Pareto optimality, and Pareto front, which have already been introduced in Chapter 1 and discussed in the previous section, in a more formal manner.

5.2.1 PARETO DOMINANCE

Given two vectors of decision variables $\mathbf{x}, \mathbf{y} \in \mathbb{R}^n$, $\mathbf{x}$ is said to Pareto dominate $\mathbf{y}$ (denoted as $\mathbf{f}(\mathbf{x}) \succeq \mathbf{f}(\mathbf{y})$; assuming a maximization problem here), if and only if $\mathbf{f}(\mathbf{x})$ is partially greater than $\mathbf{f}(\mathbf{y})$:

$$\forall i \in \{1, \ldots, k\} \ f_i(\mathbf{x}) \geq f_i(\mathbf{y}) \ \wedge \ \exists \ \forall i \in \{1, \ldots, k\} : \ f_i(\mathbf{x}) > f_i(\mathbf{y}). \tag{5.2}$$

It naturally follows from the above definition that a vector of decision variables $\mathbf{x}$ is non-dominated in region $\mathcal{H} \subset \mathbb{R}^n$, if there does not exist another vector of decision variables $\mathbf{y} \in \mathcal{H}$ such that $\mathbf{f}(\mathbf{y}) \succeq \mathbf{f}(\mathbf{x})$.

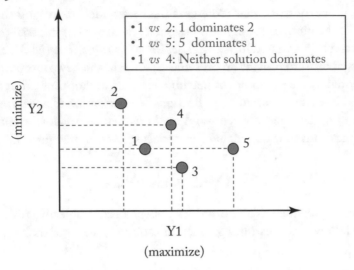

Figure 5.2: A schematic demonstrating the concept of *dominance* for a bi-objective problem (maximize Y1 and minimize Y2).

For instance, consider Figure 5.2.[1] We consider i_1 to dominate i_2, if the solution i_1 is no worse than i_2 in all objectives and if the solution i_1 is strictly better than i_2 in at least one target objective. Given a set of solutions, the *non-dominated solution set* is a set of all the solutions that are not dominated by any member of the solution set.

5.2.2   PARETO OPTIMALITY

Once the non-dominance has been defined, Pareto optimality can be formulated as: A vector of decision variables $\mathbf{x}^* \in \mathcal{F} \subseteq \mathcal{S} \subseteq \mathbb{R}^n$ is Pareto optimal if it is non-dominated with respect to $\mathcal{F}$. In other words, $\mathbf{x}^*$ is said to be Pareto optimal in $\mathcal{F}$ if there exists no other feasible vector $\mathbf{x}$ of decision variable which would increase some objective without causing a simultaneous decrease in at least one other objective.

5.2.3   PARETO FRONT

Note that the above definition of Pareto optimality does not always provide a single solution as an optimal point in the decision variable space and most frequently results in a set of solutions which are collectively referred to as the Pareto Optimal Set $\mathcal{P}^*$ defined as:

$$\mathcal{P}^* = \{\mathbf{x} \in \mathcal{F} |\ \mathbf{x} \text{ is Pareto optimal}\}. \tag{5.3}$$

[1]Some of the schematics are inspired by the lecture notes of Prof. Sudhoff (https://engineering.purdue.edu/~su dhoff/ee630/Lecture09.pdf).

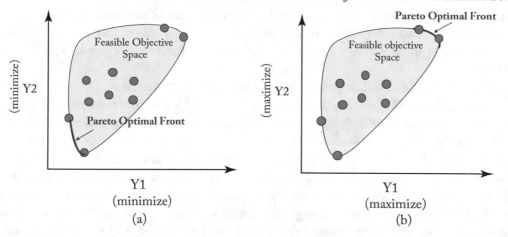

Figure 5.3: A schematic illustrating the Pareto optimal fronts for a bi-objective optimization problem. (a) The goal is to minimize both objectives (Y1 and Y2). (b) The goal is to maximize both objectives (Y1 and Y2). In both plots, the Pareto optimal front (thick red curve) forms the boundary that provides the optimal trade-off between the two properties within the feasible objective space, i.e., the area inside the thin black line.

When all elements of a Pareto Optimal Set are plotted in the objective function space, the resulting front is known as the Pareto front. One of the major goals in MOO, therefore, is determining elements of $\mathcal{P}^*$ in the allowed and physically relevant space $\mathcal{F}$ of all the decision variable vectors and the set of solutions thus arrived at represents the best possible trade-offs among the objectives for the targeted optimization problem. In Figure 5.3, the Pareto optimal fronts for the two bi-objective problems are schematically shown: (i) minimize *both* Y1 and Y2 (Figure 5.3a) and (ii) maximize *both* Y1 and Y2 (Figure 5.3b).

5.3 DESIGNING MATERIALS USING MULTI-OBJECTIVE OPTIMIZATION STRATEGIES

In the materials science literature there are at least two approaches toward MOO to accelerate the search and discovery of new materials. In the first approach, data-driven machine learning models are developed based on available training data that maps the relationship between input descriptors and target properties of interests. These models can also be referred to as *forward models* or the approach can be referred to as *forward design*. These forward models, in turn, are coupled with an optimization methods inspired by evolution and natural selection such as generic algorithms, simulated annealing, or Monte Carlo-based methods, which will survey the response surface of the trained machine learning models in the feasible design space (in the multivariate input space) and identify the Pareto optimal front in the feasible objective space (which is depicted in Figure 5.3) as captured by the modalities of the machine learning models. The use of optimization methods to identify potentially new materials based on the trained machine learning models is some times referred to as *inverse design* or *inverse modeling* in the literature.

In contrast, the second approach works a little differently and is rooted in the framework of Bayesian optimization (discussed in more detail for a single objective problem in Section 4.2). Here, data-driven machine learning models are built based on available training data that maps the relationship between input descriptors and target properties of interests. This is similar to the forward modeling approach. However, an important distinction is that these forward models *must* also have the capability to quantify prediction uncertainties.[2] The trained models are then used to predict the responses, along with the associated uncertainties, of the yet-to-be explored data points (complement of the training set). Various acquisition or utility functions are utilized (including the expected improvement that is discussed in Section 4.2) to select optimal data points from the yet-to-be explored dataset for validation and feedback. Experiments or computations are performed to provide the ground truth, which is then augmented to the training dataset for model update and next round of predictions. The iterative process is repeated until the Pareto front is established. Thus, one of the main differences is that the first approach exploits the machine learning models whereas the second approach uses exploitation-exploration tradeoff to rapidly determine the Pareto front. We will discuss both approaches in depth.

5.3.1   MULTI-OBJECTIVE OPTIMIZATION BASED ON FORWARD AND INVERSE MODELING

One of the biggest challenge toward novel materials design and discovery is that accurate property predictions/measurements for a wide range of materials properties are always laborious and resource intensive. However, given an initial set of relevant data for a target property of interest, statistical learning based surrogate model development provides an efficient alternative route toward accurate estimation of the target property for any new materials cases to be explored without needing to resort to expensive measurements or first principles based computations. As shown in Figure 5.4 and discussed in details in Chapter 2, this data-enabled forward modeling approach[3] relies heavily on finding a suitable numerical representations or *fingerprints* for the materials of interest, which are subsequently used to establish a validates and predictive structure—property mapping via a surrogate model building exercise.

Once trained and validated surrogate property prediction models are available for all properties of interest, one can, in principle, enumerate all potentially possible materials, predict target properties and visualize various tradeoffs in the underlying objective function space to select the Pareto optimal candidates. However, such a brute-force enumeration scheme can quickly be-

---

[2]Note that the uncertainty quantification capability in the machine learning model is not critical for the first approach.
[3]The approach that starts with a specific material and aims to predict its properties is conventionally referred to as the forward modeling approach.

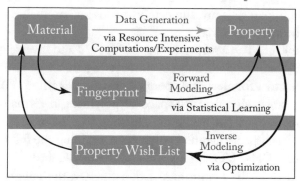

Figure 5.4: A schematic illustration of forward and inverse modeling approaches for MOO surrogate-based problems in materials science, compared with the conventional materials to property routes based on case-by-case computations or experimental measurements. While the forward modeling relies on surrogate model development for each property of interest followed by an explicit enumeration of all combinatorial possibilities to identify optimal candidate choices, the inverse modeling utilizes the surrogate models in combination with a selected optimization algorithm to pick optimal materials that meet certain objectives pre-specified in a *property wish list*.

come grossly inefficient as the number of potential candidates to be evaluated within the targeted compositional/configurational space grow rapidly. In such vast chemical space exploration problems, it is common to employ the developed surrogate property prediction models in combination with well-established optimization algorithms such as evolutionary algorithms [2], Monte Carlo methods [3], or simulated annealing [11] to sample chemistries residing on or near the targeted Pareto optimal front. In contrast to the forward modeling approach, this strategy starts with a target combination of desired properties and focuses on utilizing the adopted optimization scheme to find materials that would meet the requirements and therefore referred to as the inverse modeling approach.

Although many scientific discoveries and technological breakthroughs in the past were credited to seemingly *trial-and-error* practices and serendipity, one can easily imagine that as the search space and the dimensionality of the objective space grows the expected success rate of a random exploration would rapidly shrink. On the other hand, a rational design approach based on accurate and efficient surrogate property models can be much more effective and provide a systematic way to improve. Of course, an approach would only work if it is possible to explicitly state the initial screening criteria in terms of calculable/measurable properties and machine learning based property prediction models can be developed for the properties. Furthermore, for practical reasons, a hierarchical step-wise search approach is often preferred where instead of focusing on all objectives simultaneously, a selected subset of the most critical ones is first considered and identified materials are subjected to subsequent analysis and testing for other relevant attributes in the search criteria.

A practical example of MOO surrogate-based forward and inverse modeling was recently presented by Mannodi-Kanakkithodi and co-workers [2] that was aimed at design of next-generation polymer dielectrics for electrical energy storage applications. Focusing on linear polymers, they defined the target chemical subspace of polymers in terms of chemical structural units (or polymer motifs forming the backbone), leading to a list of combinatorial possibilities. Specifically, the chemical subspace included polymers containing the following seven commonly occurring building blocks: CH_2, CO, NH, C_6H_4, C_4H_2S, CS, and O.

As briefly alluded to in the beginning of this chapter, the energy stored in a dielectric capacitor is directly proportional to ϵE_b^2, where ϵ represents the dielectric constant of the materials and E_b is breakdown field strength—the maximum electric field the material can withstand without failing. Therefore, the dielectric design exploration considered both the relevant attributes. However, given that the explicit computations of E_b in materials can be highly challenging, the electronic bandgap (which is known to correlate well with E_b and can be computed reliably with electronic structure computations) was used as a proxy for the E_b. Out of the vast possible polymer chemical universe within the selected seven commonly occurring building blocks, a controlled sub-space was selected to generate training data using high-throughput computations. For this all physically meaningful and synthetically feasible four-motif polymer compositions and configurations that can be formed to represent the polymer chain repeat unit were enumerated. Next, for these selected cases density functional perturbation theory [12] was used to compute the dielectric constants and bandgaps were computed using the hybrid Heyd–Scuseria–Ernzerhof (HSE06) electronic exchange-correlation functional [13], which is known to fix the problem of bandgap underestimation associated with conventional DFT to a large extent. Within the adopted framework, computed dielectric constants and bandgaps match very well with experimentally measured results for inorganic compounds as well as common polymers [14].

As a next step, the generated computational training dataset was employed to develop machine learning-based predictive models for the dielectric constant and the bandgap. The electronic and lattice components of the dielectric constant (namely, the electronic and ionic dielectric constants, respectively) were modeled separately owing to the different underlying mechanistic origins of the two properties form a physical perspective. Use of chemo-structural fingerprints—explicitly accounting for compositions of various motifs and their relative arrangements in the polymer backbone—along with the kernel ridge regression machine learning algorithm lead to accurate surrogate prediction models for each of the three properties, as shown in Figure 5.5a–c. After rigorous validation of the machine learning predictions against DFT-based computations and experimental measurements (as shown in Figure 5.5d–g), the developed models models were used to make predictions over a much larger part of the chemical space, going beyond the four-block polymer dataset which was originally used to train these models.

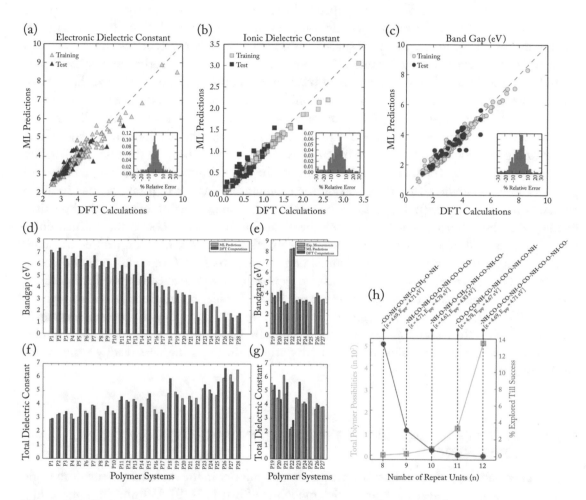

Figure 5.5: Parity plots comparing training and test set prediction performances of machine learning-based surrogate models against computationally demanding electronic structure calculations for (a) electronic and (b) ionic dielectric constants and (c) bandgaps of organic polymer dielectrics. The insets show the test set prediction error distributions in each case. Further validation of machine learning predictions with electronic structure calculations and experimental measurements for (d, e) bandgap and (f, g) dielectric constant of several new polymers not included in the machine learning training dataset. (h) Performance advantage of using surrogate-based inverse modeling approach as combinatorial possibilities in the target chemical space rise (see test for details). Adapted from Ref. [2], with permission.

For the forward modeling, all symmetry unique n-block polymers (for specific values of n with $n > 4$) were enumerated and fingerprinted to make predictions directly using the trained learning models. For the inverse modeling, on the other hand, an evolutionary algorithm was used to explore the polymer chemical space in search of a candidate polymer with pre-specified bandgap and dielectric constant values. A comparative analysis of both the approaches is presented in Figure 5.5h demonstrating the advantage of the inverse approach when faced with a combinatorial explosions in materials possibilities to be explored. Starting with the 8-block polymers, where there are $\sim$150,000 total material possibilities, the evolutionary algorithm is able to traverse a small percentage of the points in determining the required polymer(s). Upon going to higher block systems, like 9-block or 10-block polymers, while the total possibilities rise exponentially, the fraction of total possibilities the algorithm needs to explore in each case gets smaller and smaller. Therefore, despite the exponential increase in total polymer possibilities, as the number of repeating units n increases, only a small fraction of the overall possible candidate materials is actually evaluated by the algorithm in order to obtain the optimal polymer(s). Figure 5.5h also presents the identified n-block polymers obtained for different values of n for a target dielectric constant of 5 and a target bandgap of 5 eV. In addition to the size of the chemical space, the advantage of the inverse approach can be expected to grow further as the number of dimensions in the objective space get larger. Similar conclusions were also reached within a Monte Carlo-based search study [3]. Finally, we note that while the presented example does not specifically account for constraints, in addition to the optimization objectives, implementation of constraint-handling techniques in optimization algorithm are readily available for MOO problems [15, 16].

5.3.2   MULTI-OBJECTIVE OPTIMIZATION BASED ON OPTIMAL LEARNING

Approaches based on optimal learning are gaining popularity in the materials science and engineering community to accelerate predictive modeling efforts [17–19]. Within this approach, a data-driven model becomes the basis for predicting new and previously unexplored data point(s), which is (are) then fed into an "oracle" (a high-fidelity computational code or a series of experiments) that will evaluate the true response and provide feedback for model update. Typically, the whole process of prediction, validation, and feedback is repeated until the budget runs out or sufficiently improved designs are achieved (that is, new materials with desired properties are discovered). The update protocol ensures that the model is fed with validated data that will allow for improvement. It is important to note that the original model can be relatively poor (that is, can have large mean squared error), but the iterative feedback loop sequentially improves the performance. In Section 4.2, the concept of Bayesian optimization was discussed in depth. As a result, we will only focus on salient features that are relevant to coherently explain the ideas behind MOO design.

Consider a single objective optimization problem of finding a minimum in the target property of interest. Let Y be denoted as the property of interest that is dependent on a set of descriptors X. Let N denote the number of data points in the training set. The role of machine learning methods is then to establish a function $\hat{Y} = f(X)$ from the training data such that a loss function, say $\frac{1}{N}\sum_{i=1}^{N}(Y - \hat{Y})^2$, is minimized. When the goal is to find the smallest y, we can choose a newly calculated design point $Y(X^{N+1})$ representing an improvement over the current best design, $f^{min}(X) = argmin[f^1(X^{(1)}), f^2(X^{(2)}), \ldots, f^N(X^{(N)})]$ using utility functions such as $P[I]$ and $E[I]$, the probability and expected value of improvement, respectively. Before we discuss $P[I]$ and $E[I]$, the improvement I can be defined as

$$I = f^{min}(X) - \hat{Y}\left(X^{(N+1)}\right). \tag{5.4}$$

Then, the $P[I]$ becomes

$$P[I] = P\left[Y\left(X^{(N+1)}\right) \leqslant f^{min}(X)\right]$$

$$= \int_{-\infty}^{f^{min}(X)} \frac{1}{\sigma(X^{(N+1)})\sqrt{(2\pi)}} \exp\left(-\frac{[\hat{Y} - \mu(X^{(N+1)})]^2}{\sigma^2(X^{(N+1)})}\right) \tag{5.5}$$

$$= \Phi\left[\frac{f^{min}(X) - \mu(X^{(N+1)})}{\sigma(X^{(N+1)})}\right].$$

If P is a Gaussian distribution, the function Φ is the cumulative distribution function of the Gaussian integrand. Similarly, it can be shown that the expected improvement, $E[I]$ can be defined as

$$E[I] = \left[f^{min}(X) - \hat{Y}\left(X^{(N+1)}\right)\right] \Phi\left[\frac{f^{min}(X) - \mu(X^{(N+1)})}{\sigma(X^{(N+1)})}\right]$$

$$+ \sigma\phi\left[\frac{f^{min}(X) - \mu(X^{(N+1)})}{\sigma(X^{(N+1)})}\right], \tag{5.6}$$

where ϕ is the Gaussian probability density function.

Now, consider a two-objective optimization problem. The green shaded region in Figure 5.6 indicates the region where the occurrence of a candidate material after measurement would result in an improvement over the current front shown in blue dots. This region means that the current sub-pareto front would be modified to include the newly measured material. The probability of improvement $P[I]$ that the new point is an improvement over all existing points is the total probability of a candidate data-point integrated over the green shaded region in Figure 5.6 and is given by

$$P[I] = \int_{Shaded} \phi(Y_1, Y_2) dY_1 dY_2, \tag{5.7}$$

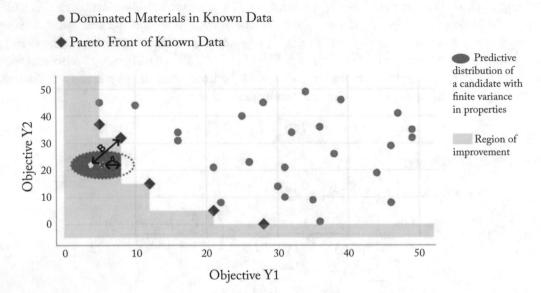

Figure 5.6: A schematic representation of data and its Pareto front for a bi-objecitve optimization problem, where Y1 and Y2 are the two target properties and both the properties are to be minimized. The Pareto front will be convex toward the origin if all the properties were to be maximized. For a mixed problem with both minimization and maximization, the concaveness of the Pareto front will be rotated by 90°. The diamond points in red color represent the Pareto front of data whereas the blue color dots are the points which are dominated by the Pareto front. The region in white background is the dominated region and the green shaded region is the region of improvement. Occurrence of a new material in the green shaded region could replace at least one existing Pareto front point and thus lead to an improvement from the current Pareto front. The brown shaded area corresponds to the predicted distribution of one candidate material. The violet point inside the brown shaded region represents the mean value of the predicted distribution. The yellow point indicates the centroid of the predicted distribution lying inside the region of improvement. It is possible that the entire predicted distribution of some candidate material may lie inside the region of improvement. In that case, the mean of the entire distribution would coincide with the centroid. The distances A and B represent $L_{maximin}$ and $L_{centroid}$, respectively. Adapted from [4].

where Y_1 and Y_2 are the target objectives and $\phi(Y_1, Y_2)$ is the uncorrelated Gaussian probability distribution function formed from the mean and variance of Y_1 and Y_2 distributions with $\phi(Y_1, Y_2) = \phi(Y_1)\phi(Y_2)$. We have therefore assumed a Gaussian distribution for the predicted values with a mean and variance. Similarly, the equivalent two objective expected improvement $E[I(x)]$ is the first moment of I of the joint probability distribution $\phi(Y_1, Y_2)$ over the green area in Figure 5.6 about the current sub-pareto front. Geometrically, we can calculate $E[I(X)] = P[I(X)]L$ in two ways depending on how the "length" L is evaluated.

- Centroid approach to $E[I]$, referred to as EI-Centroid: $E[I(X)] = P[I(X)]L$, where

$$L = \sqrt{(C_1(X) - Y_1(X))^2 + (C_2(X) - Y_2(X))^2},$$

the distance between the centroid $(C_1(X), C_2(X))$ at the candidate data point, X, and closest point on the sub-pareto front, $(Y_1(X), Y_2(X))$. The centroid of the probability distribution for the candidate point in the green shaded region is calculated using

$$C_1(X) = \int_{Shaded} Y_1\phi(Y_1, Y_2)dY_2 dY_1 / P[I]. \tag{5.8}$$

Similarly for $C_2(X)$.

- Maximin approach to $E[I]$, referred to as EI-maximin: Let the mean predicted values for a candidate material be (μ_1, μ_2). Then we define the distance $d_{maximin} = Max_i(Min(p_{i1} - \mu_1, p_{i2} - \mu_2), 0)$, where $P_i = (p_{i1}, p_{i2})$ and $P_i \in$ Pareto front. The EI-maximin is then

$$EI_{maximin} = d_{maximin} \times P[I(X)].$$

Thus, for each candidate point in the region of improvement, EI-centroid is calculated by taking the product of $P[I]$ with the minimum distance between points on the known sub pareto front and centroid of the probability distribution within the region of improvement. The candidate point with the largest EI-centroid is then the choice for the next measurement. EI-maximin is the product of $P[I]$ and the maximum of the minimum distance of either of the means (μ_1, μ_2) of a particular candidate point from individual sub Pareto front points p_i. The former considers improvement over the properties Y_1, Y_2 combined, whereas EI-maximin considers each property separately, takes the one which is smaller from a particular sub Pareto point, and then then maximizes that among all the sub Pareto points. Both strategies select a datapoint such that its measurement produces maximum modification to the sub-Pareto front.

Algorithm

Algorithm 5.1, in pseudo code form, given the data, surrogate, and choice of $E[I]$, is given below.

Algorithm 5.1 Multi-objective design algorithm based on optimal learning

---

1: **input1:** X_{train} ← list containing descriptor vectors of all known materials
2: **input2:** Y_{train} ← list containing property values vectors of all known materials
3: **input3:** X_{search} ← list containing descriptor vectors of all unmeasured materials
4: **procedure** DESIGN($X_{train}, Y_{train}, X_{search}$)
5:      Build a surrogate model, $f(X_{train}) = Y_{train}$
6:      **for all** descriptor vectors $x_{search}^i \in X_{search}$ **do**
7:          Bootstrap the predictions $f(x_{search}^i) \to g(x_{search}^i)$
8:          Mean value of the predicted distribution, $\mu_{x_{search}^i} \leftarrow E[g(x_{search}^i)]$
9:          Uncertainty in the predicted distribution, $\sigma_{x_{search}^i} \leftarrow \sqrt{E[(g(x_{search}^i) - \mu_{x_{search}^i})^2]}$
10:         Calculate the Probability of Improvement, $P[I]_{x_{search}^i} = P[g(x_{search}^i) \in$ Region of Improvement]
11:         Calculate Improvement, I. $I_{x_{search}^i} = Maximin(|\mu_{x_{search}^i} - PF|)$ **or** $I_{x_{search}^i} = Centroid(|\mu_{x_{search}^i} - PF|)$
12:         Expected Improvement, $E[I]_{x_{search}^i} = I_{x_{search}^i} \times P[I]_{x_{search}^i}$
13:     **end for**
14:     $x_{selected} = x_{search}^i \in X_{search}$ such that $E[I]_{x_{search}^i} > E[I]_{x_{search}^i} \; \forall \; x_{search}^j \in X_{search}, i \neq j$
15:     **return** $x_{selected}$
16: **end procedure**

Materials Application

The dataset that we have chosen is from the work of Xue et al. [20], where the solid solutions belonging to the multicomponent NiTi-based alloys, where chemical compositions can be generally described as $Ti_{50}Ni_{1-x-y-z}Cu_xFe_yPd_z$ with the targeted property of low thermal hysteresis or dissipation were synthesized. These are promising shape memory alloys, whose functionalities include the shape memory effect and superelasticity that arise from the reversible martensitic transformation between high temperature austenite and low temperature martensite phases. Heating and cooling across the martensitic transformation temperature results in hysteresis as the transformation temperatures do not coincide, giving rise to fatigue. Only a single objective, thermal hysteresis, was previously predicted and all the alloys constrained by 50-x-y-z $\geq$ 30%, x $\leq$ 20%, y $\leq$ 5%, and z $\leq$ 20% were synthesized by the same processing protocols in the same materials processing laboratory. The goal is to find alloy compositions in the dataset that minimizes *both* the thermal hysteresis and the transition temperature.

Each alloy is described in terms of one or more features representing aspects of structure, chemistry, bonding. While there are many approaches to choosing features, this work used prior materials knowledge to construct the features. It is known that the martensitic transition temperatures, which affect thermal hysteresis, are strongly correlated with the valence electron concentration and electron number per atom. In particular, the martensite and austenite start temperatures vary significantly when the valence electron concentration increases and show behavior that depends on the electron valence number per atom. Moreover, the thermal hysteresis is directly influenced by the atomic size of the alloying elements as the hysteresis increases with size at almost constant electron valence number. Thus, Zunger's pseudopotential radii [21], Pauling electronegativity, metallic radius, valence electron number, Clementi's atomic radii [22], and Pettifor chemical scale [23] were used as features for building the machine learning model [20].

The results from the MOO design for this dataset is shown in Figure 5.7. From Figure 5.7a, it is clear that employing MOO design strategies decreases the number of measurements required to find the optimal Pareto front by nearly 20% compared to random selection. In addition, the MOO strategies reduced the computational effort by 40–45% compared to employing brute-force search to calculate all the candidate materials. The EI-centroid based design strategy and pure exploration perform similarly well; however, the EI-maximin approach shows superior performance compared to all other strategies, particularly if the initial training dataset sizes are small. As shown in Figure 5.7b, there are seven points in the optimal Pareto front for this data set. The iterative design was carried out by varying the training set sizes from 5–70 (*x*-axis).

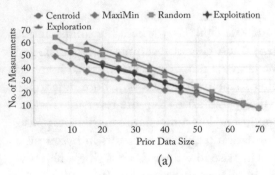

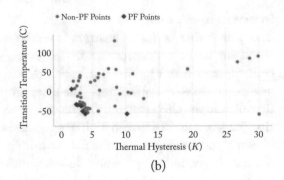

(a)   (b)

Figure 5.7: (a) The size of the initial training dataset (at the 0th iteration) is plotted against the average number of design cycles required to find all the points in optimal Pareto front. When the size of initial training data is small, not surprisingly the regression models produce a less-reliable fit to the data. Thus, the EI-maximin design strategy in which the exploration and exploitation of data is more balanced, performs much better than all other methods. (b) Optimal Pareto front for the dataset used in (a). There are seven points in the optimal Pareto front for this dataset. The optimal Pareto front is considered as unknown at the beginning of design process. Starting randomly from a set of training data, the goal is to find all the optimal Pareto front points in as few design cycles as possible. The red colored square points form the optimal Pareto front whereas each blue colored point is dominated by at least one point in the optimal Pareto front. Adapted from [4].

5.4   REFERENCES

[1] Z. Li, K. G. Pradeep, Y. Deng, D. Raabe, and C. C. Tasan. Metastable high-entropy dual-phase alloys overcome the strength—ductility trade-off. *Nature*, 534:227, 2016. DOI: 10.1038/nature17981 117

[2] A. Mannodi-Kanakkithodi, G. Pilania, T. D. Huan, T. Lookman, and R. Ramprasad. Machine learning strategy for accelerated design of polymer dielectrics. *Scientific Reports*, 6:20952, 2016. DOI: 10.1038/srep20952 117, 123, 124, 125

[3] A. Mannodi-Kanakkithodi, G. Pilania, R. Ramprasad, T. Lookman, and J. E. Gubernatis. Multi-objective optimization techniques to design the Pareto front of organic dielectric polymers. *Computational Materials Science*, 125:92, 2016. DOI: 10.1016/j.commatsci.2016.08.018 117, 123, 126

[4] A. M. Gopakumar, P. V. Balachandran, D. Xue, J. E. Gubernatis, and T. Lookman. Multi-objective optimization for materials discovery via adaptive design. *Scientific Reports*, 8(1):3738, 2018. DOI: 10.1038/s41598-018-21936-3 118, 128, 132

[5] C. A. Coello and R. L. Becerra. Evolutionary multiobjective optimization in materials science and engineering. *Materials and Manufacturing Processes*, 24(2):119, 2009. DOI: 10.1080/10426910802609110 119

[6] S. Ganguly, S. Datta, and N. Chakraborti. Genetic algorithms in optimization of strength and ductility of low-carbon steels. *Materials and Manufacturing Processes*, 22(5):650, 2007. DOI: 10.1080/10426910701323607

[7] W. Paszkowicz. Genetic algorithms, a nature-inspired tool: Survey of applications in materials science and related fields. *Materials and Manufacturing Processes*, 24(2):174, 2009. DOI: 10.1080/10426910802612270.

[8] J. Knowles. ParEGO: A hybrid algorithm with on-line landscape approximation for expensive multiobjective optimization problems. *IEEE Transactions on Evolutionary Computation*, 10(1):50, 2006. DOI: 10.1109/tevc.2005.851274

[9] A. Agarwal, F. Pettersson, A. Singh, C. S. Kong, H. Saxén, K. Rajan, S. Iwata, and N. Chakraborti. Identification and optimization of AB_2 phases using principal component analysis, evolutionary neural nets, and multiobjective genetic algorithms. *Materials and Manufacturing Processes*, 24(3):274, 2009. DOI: 10.1080/10426910802678321

[10] B. J. Reardon and S. R. Bingert. Inversion of tantalum micromechanical powder consolidation and sintering models using Bayesian inference and genetic algorithms. *Acta Materialia*, 48(3):647, 2000. DOI: 10.1016/s1359-6454(99)00407-3 119

[11] T. D. Huan, A. Mannodi-Kanakkithodi, and R. Ramprasad. Accelerated materials property predictions and design using motif-based fingerprints. *Physical Review B*, 92:014106, 2015. DOI: 10.1103/physrevb.92.014106 123

[12] S. Baroni, S. De Gironcoli, A. Dal Corso, and P. Giannozzi. Phonons and related crystal properties from density-functional perturbation theory. *Reviews of Modern Physics*, 73(2):515, 2001. DOI: 10.1103/revmodphys.73.515 124

[13] J. Heyd, G. E. Scuseria, and M. Ernzerhof. Hybrid functionals based on a screened Coulomb potential. *Journal of Chemical Physics*, 118(18):8207, 2003. DOI: 10.1063/1.1564060 124

[14] T. D. Huan, A. Mannodi-Kanakkithodi, C. Kim, V. Sharma, G. Pilania, and R. Ramprasad. A polymer dataset for accelerated property prediction and design. *Scientific Data*, 3:160012, 2016. DOI: 10.1038/sdata.2016.12 124

[15] C. A. Coello. Theoretical and numerical constraint-handling techniques used with evolutionary algorithms: A survey of the state of the art. *Computer Methods in Applied Mechanics and Engineering*, 191(11):1245, 2002. DOI: 10.1016/s0045-7825(01)00323-1 126

[16] E. Mezura-Montes. *Constraint-Handling in Evolutionary Optimization*, volume 198, Springer, Heidelberg, 2009. DOI: 10.1007/978-3-642-00619-7 126

[17] A. J. Keane. Statistical improvement criteria for use in multiobjective design optimization. *AIAA Journal*, 44(4):879, 2006. DOI: 10.2514/1.16875 126

[18] J. Svenson and T. Santner. Multiobjective optimization of expensive-to-evaluate deterministic computer simulator models. *Computational Statistics and Data Analysis*, 94:250, 2016. DOI: 10.1016/j.csda.2015.08.011

[19] I. Voutchkov and A. Keane. Multi-objective optimization using surrogates. In Y. Tenne and C.-K. Goh, Eds., *Computational Intelligence in Optimization: Applications and Implementations*, p. 155, Springer, Heidelberg, 2010. DOI: 10.1007/978-3-642-12775-5_7 126

[20] D. Xue, P. V. Balachandran, J. Hogden, J. Theiler, D. Xue, and T. Lookman. Accelerated search for materials with targeted properties by adaptive design. *Nature Communications*, 7:11241, April 2016. DOI: 10.1038/ncomms11241 131

[21] A. Zunger. Systematization of the stable crystal structure of all AB-type binary compounds: A pseudopotential orbital-radii approach. *Physical Review B*, 22:5839, 1980. DOI: 10.1103/physrevb.22.5839 131

[22] E. Clementi, D. L. Raimondi, and W. P. Reinhardt. Atomic screening constants from SCF functions. II. Atoms with 37 to 86 electrons. *Journal of Chemical Physics*, 47(4):1300, 1967. DOI: 10.1063/1.1712084 131

[23] D. G. Pettifor. A chemical scale for crystal-structure maps. *Solid State Communications*, 51(1):31, 1984. DOI: 10.1016/0038-1098(84)90765-8 131

CHAPTER 6

Multi-Fidelity Learning

The title of one of the seminal papers in the field of *multi-fidelity learning and optimization*, "Predicting the Output from a Complex Computer Code when Fast Approximations are Available," gives a good hint at what multi-fidelity optimization and learning is. It is about building an accurate surrogate model for a complex problem based on the partial use of data obtained from calculations that are less accurate than what is possible. The methods need not involve only calculations. They can combine results from experiments performed with different levels of precision. The objective of multi-fidelity optimization is to reduce the need of performing many accurate and presumably costly calculations to scope complex problems by performing fewer such calculations but supplementing them with less costly ones. In materials research, an obvious application is the computation of bandgaps. There is a known hierarchy of choices of pseudo-potentials that range from those lending themselves to calculations quickly and cheaply performed but less accurate in their predictions of bandgaps to those that are more time-consuming and expensive to use but more accurate in their predictions. In multi-fidelity optimization, a Gaussian process is typically called *kriging* after D. G. Krige, a South African mining engineer who pioneered the method in making mineral distribution predictions based on core samples [1]. Multi-fidelity learning is often called *co-kriging*.

6.1   KRIGING

In this section, we summarize the basic results of kriging (Gaussian process regression). More details about Gaussian processes and the related BGO, which uses Gaussian processes, are given in Appendix C. Then, in the second section, we summarize the co-kriging regression method as an extension of the kriging method to situations where model outputs of different levels of fidelity, that is, variations both in computational cost and accuracy, are available to *learn* a property of interest. The approach summarized was established by Kennedy and O'Hagan [2] who put forward a coherent mathematical framework for blending heterogeneous variable-fidelity information sources, creating a natural setting for multi-fidelity modeling. Subsequently, Forrester et al. [3] adopted this framework to build a two-level co-kriging scheme, discussed the details of the estimation of model hyper-parameters, and successfully demonstrated its application.

Kriging, that is, Gaussian process regression, is a form of Bayesian inference. As a Bayesian method (Chapter 7 and Appendix A), it starts with a prior distribution over functions. This prior has the form of a Gaussian process in the sense that its samples are normally distributed and the covariance between any two samples is a covariance function (kernel) evaluated at the locations of the two samples. For a set of observed values, each value is associated with a location. By combining the Gaussian prior with a Gaussian likelihood function for each of the observed values, we can predict new values at any new location. The resulting posterior distribution is also Gaussian, with a mean and covariance that are easily computed from the observed values, their variance, and the kernel matrix derived from the prior.

Suppose we have a training set of n observations $\mathcal{D} = \{(\boldsymbol{x}_i, y_i), |i = 1, \ldots, n\}$, where an $\boldsymbol{x}_i$ is a vector of D features. Alternately, we can write this as $\mathcal{D} = (X, \boldsymbol{y})$, where X is a $D \times n$ matrix whose columns are the feature vectors and $\boldsymbol{y}$ is a vector collecting the labels (that is, the targets) y_i. What we seek in the kriging method is the conditional probability of the targets given the inputs. Even more specifically, if $\boldsymbol{x}_*$ is a new input, we seek $p(y_*|\boldsymbol{x}_*, \mathcal{D})$, the probability distribution of the value of a new target y_* given a proposed x_* and the data $\mathcal{D}$.

As the kriging method is based on Gaussian processes, our objective function $f(\boldsymbol{x})$ is regarded as something sampled from a Gaussian defined by a mean $\mathbf{m}(\boldsymbol{x})$ and a covariance matrix. If the expectation of the mean is chosen to be zero, leaving $\mathrm{cov}(f(\boldsymbol{x}), f(\boldsymbol{x}')) = k(\boldsymbol{x}, \boldsymbol{x}')$, where $k(\boldsymbol{x}, \boldsymbol{x}')$ is some kernel function, we have

$$
\begin{bmatrix} \boldsymbol{y} \\ f_* \end{bmatrix} = \mathcal{N}\left(\begin{bmatrix} 0 \\ 0 \end{bmatrix} \middle| \begin{bmatrix} 0 \\ 0 \end{bmatrix}, \begin{bmatrix} K(X, X) + \sigma_n^2 I & K(X, X_*) \\ K(X_*, X) & K(X_*, X_*) \end{bmatrix}\right), \tag{6.1}
$$

where $K(X, X_*)$ is the $n \times 1$ matrix of covariance pairs between the training data and test point. $K(X, X)$ is an $n \times n$ matrix, $K(X_*, X)$ is an $1 \times n$ matrix and $K(X_*, X_*)$ is a 1×1 matrix similarly defined. With the use of the properties of multivariate Gaussians and Bayes's Theorem, our predictive distribution becomes

$$
p(f_*|\mathbf{x}, \mathcal{D}) = p(f_*|\boldsymbol{x}_*, X, \boldsymbol{y}) = \mathcal{N}(\hat{f}_*, \mathrm{cov}(\hat{f}_*)), \tag{6.2}
$$

where

$$
\begin{aligned}
\hat{f}_* &\equiv K(X_*, X)[K(X, X) - \sigma_n^2 I]^{-1} \boldsymbol{y}, \tag{6.3} \\
\mathrm{cov}(\hat{f}_*) &= K(X_*, X_*) - K(X_*, X)[K(X, X) - \sigma_n^2 I]^{-1} K(X, X_*). \tag{6.4}
\end{aligned}
$$

A convenient choice for the kernel is the *generalized squared exponential*

$$
k(\boldsymbol{x}, \boldsymbol{x}') = \sigma \exp\left(-\tfrac{1}{2} \sum_{i=1}^{D} \theta_i (x_i - x_i')^2\right).
$$

The hyperparameters, $(\sigma, \theta_1, \ldots, \theta_D)$, are what we need to learn from the data $\mathcal{D}$.

What we just described is called *simple kriging*, meaning the predictive mean is assumed to be a constant over the entire domain of the search space. There is also *ordinary kriging* which assumes the mean is constant only in the neighborhood of each data point. Other popular variants include empirical Bayesian kriging and universal kriging. *Empirical Bayesian kriging* automatically adjusts the parameters in the model and accounts for the errors in the data. *Universal kriging* accounts for local trends in the data by assuming a position dependent predictive mean. Here, typically one takes $\mu(\mathbf{x}) = \sum_i \beta_i b_i(\mathbf{x})$, where the *trend parameters* β_i are an additional set of hyperparameters to be determined and the $b_i(\mathbf{x})$ are a user-specified set of functions.

Some points to note about kriging are that the kernel trick (Appendix C) allows us to extend the kernel function, whose hyperparameters are fitted to the known data, to express correlations with this data and the proposed features $\mathbf{x}_*$ and the subsequent block matrix structure of the covariance matrix is readily accommodated as a multivariate Gaussian for such tasks as marginalization. Kriging works with data of a single fidelity source. The resultant block structure is $(1 + 1) \times (1 + 1)$.

6.2   CO-KRIGING

While the auto-regressive scheme of Kennedy and O'Hagan [2] is quite general, in the sense that we can apply it to a situation where s-levels of variable-fidelity estimates are available, it suffers from practical limitations pertaining to computational efficiency when the number of levels s or number of data points grow large. A breakthrough was achieved by recent work of Le Gratiet and Garnier [4, 5] who proved that any co-kriging scheme with s-levels of variable-fidelity information sources decouples and equivalently reformulates in a recursive fashion as s kriging problems. This result allows the construction of predictive co-kriging schemes by solving a sequence of simpler kriging problems.

Co-kriging uses information from several variable types, but this use comes at a price. Let us say that the main variable of interest is $f(X)$. Co-kriging requires estimating the autocorrelations for each variable as well as all cross-correlations between $f(X)$ and all other variable types. These correlations are used to make better predictions. Theoretically, with co-kriging you can do no worse than kriging for $f(X)$ because if there is no cross-correlation, you fall back onto the autocorrelations for $f(X)$, but if you look at the variations in the differences of two or more processes, you might do worse than looking at any individual processes.

To motivate the advantages of using information from multiple variable types, we now consider simple co-kriging, constant mean in the neighborhood of a point, for the following two models:

$$\begin{aligned}
f(X) &= F + \varepsilon_1(X) \\
g(Y) &= G + \varepsilon_2(Y),
\end{aligned}$$

where the means F and G of $f(X)$ and $g(Y)$ are unknown constants. We want to consider the case where f and g model the same process; that is, $f(X) \approx g(Y)$. The random variables X and Y need not necessarily be independent. We have two types of random errors, $\varepsilon_1(X)$ and $\varepsilon_2(Y)$, so there is an autocorrelation among the $f(X)$ and the $g(Y)$ and cross-correlation between the f and g.

Simple co-kriging attempts to predict $f(X)$, just like ordinary kriging, but it uses information in the covariate $g(Y)$, that is, the cross-correlations, in an attempt to do a better job.

How might this happen? Suppose we can compute $\mathrm{cov}(X, Y)$, want to estimate $\langle X \rangle$, know $\langle Y \rangle$, and compute $\mathrm{var}(Y)$. If we construct the random variable $X(a) = X + a(Y - \langle Y \rangle)$ and note that X and $X(a)$ have the same means, we write

$$\mathrm{var}(X(a)) = \mathrm{var}(X) - 2a\mathrm{cov}(X, Y) + a^2\mathrm{var}(Y).$$

If we take $a = \mathrm{cov}(X, Y)/\mathrm{var}(Y)$, then

$$\mathrm{var}(X(a)) = [1 - \rho_{XY}]\mathrm{var}(X) < \mathrm{var}(X).$$

Here, the correlation coefficient,

$$\rho_{XY} = \frac{(X - \langle X \rangle)(Y - \langle Y \rangle)}{\sqrt{\mathrm{var}(X)\mathrm{var}(Y)}} = \frac{\mathrm{cov}(X, Y)}{\sqrt{\mathrm{var}(X)\mathrm{var}(Y)}}$$

satisfies $-1 \leq \rho_{XY} \leq 1$.

Co-kriging is a Bayesian method based on multi-variant Gaussian processes. A key difference from kriging is a richer block matrix structure for the covariance matrix that expresses correlations among the values of variables from models with different levels of fidelity: If the number of fidelity levels is s, this block structure is $(s + 1) \times (s + 1)$.

In co-kriging, we suppose we have s training sets of n_j observations $\mathcal{D}_j = \{(x_i^{(j)}, y_i^{(j)}), | i = 1, \ldots, n\}$, $j = 1, \ldots, s$. Alternatively, we can write this as $\mathcal{D}_j = (X_j, \mathbf{y}_j)$, where X_j is a $D \times n_j$ matrix whose columns are the feature vectors and $\mathbf{y}_j$ is a vector collecting the labels (that is, the targets $y_i^{(j)}$). We assume that the precision level of the targets increases as the index j increases from 1 to s. Our data set is now $\mathcal{D} = \{\mathcal{D}_1, \mathcal{D}_2, \ldots, \mathcal{D}_s\}$. What we again seek is the conditional probability of the targets given the inputs. Even more specifically, if x_* is a new input, we seek $p(y_*|x_*, \mathcal{D})$, the probability distribution of the value of the new target y_*, given a proposed x_* and the concatenated data $\mathcal{D}$. This distribution is still a Gaussian whose mean and covariance provide the estimates for y_* and the error estimate for y_*. The predicted new target value is at the highest level of precision.

In the present work, for convenience of exposition, we discuss a two-level co-kriging model representing the high-fidelity (expensive and accurate) and low-fidelity (cheap and less trustworthy) estimates. Extension to s levels of precision is reasonably straightforward. Closely following the conventions introduced by Forrester et al. [3], we represent the n_e targets computed expensively as $\mathbf{y}_e$ for a set of materials represented by a matrix $\mathbf{X}_e$ and the n_c targets computed cheaply as $\mathbf{y}_c$ for a set of materials represented by $\mathbf{X}_c$. Further, we assume that $\mathbf{X}_e$ is a subset of $\mathbf{X}_c$; that is, $\mathbf{X}_e \subseteq \mathbf{X}_c$. This latter requirement is useful for estimating model parameters at different levels of fidelity. These datasets are concatenated to give a combined set of points as:

$$\mathbf{X} = \begin{pmatrix} \mathbf{X}_c \\ \mathbf{X}_e \end{pmatrix} = \left(x_c^{(1)}, \ldots, x_c^{(n_c)}, x_e^{(1)}, \ldots x_e^{(n_e)} \right)^{\mathrm{T}}$$

$$\mathbf{y} = \begin{pmatrix} \mathbf{y}_c \\ \mathbf{y}_e \end{pmatrix} = \left(y_c^{(1)}, \ldots, y_c^{(n_c)}, y_e^{(1)}, \ldots, y_e^{(n_e)} \right)^{\mathrm{T}}, \tag{6.5}$$

where each $\mathbf{x}_c^{(i)}$ and $\mathbf{x}_e^{(i)}$ is a d-dimensional feature vector that uniquely fingerprints the materials in a target space. Furthermore, in a close analogy with kriging, co-kriging also assumes that the value at a point in $\mathbf{X}$ is a realization of a Gaussian random variable. Therefore, we also have a random field

$$
\begin{aligned}
\mathbf{Z} &= \begin{pmatrix} \mathbf{Z}_c(\mathbf{X}_c) \\ \mathbf{Z}_e(\mathbf{X}_e) \end{pmatrix} \\
&= \left(Z_c(\mathbf{x}_c^{(1)}), \ldots, Z_c(\mathbf{x}_c^{(n_c)}), Z_e(\mathbf{x}_e^{(1)}), \ldots, Z_e(\mathbf{x}_e^{(n_e)}) \right)^{\mathrm{T}},
\end{aligned}
$$

where $Z_c(\mathbf{x})$ and $Z_e(\mathbf{x})$ are the Gaussian processes for the cheap and expensive data. If $\mathbf{x}_*$ is a proposed $\mathbf{x}$, the predicted $y_* = Z_e(\mathbf{x}_*)$. The task is to set this Gaussian process given $\mathbf{X}$ and $\mathbf{y}$. This process is set once we specify its mean and covariance. Here we use the scheme of Kennedy and O'Hagan [2].

The autoregressive scheme of Kennedy and O'Hagan [2] expresses the high-fidelity output as the low-fidelity output multiplied by a scaling factor ρ plus an another independent Gaussian process Z_d (that is, $Z_c(\mathbf{x}) \perp Z_d(\mathbf{x})$) which represents the difference between ρZ_c and Z_e:

$$
Z_e(\mathbf{x}) = \rho Z_c(\mathbf{x}) + Z_d(\mathbf{x}). \tag{6.6}
$$

This expression directly follows from the Markov-like property introduced by Kennedy and O'Hagan that assumes if given $Z_c(\mathbf{x})$, then we can learn nothing more about $Z_e(\mathbf{x})$ from any other model output $Z_c(\mathbf{x}')$, for $\mathbf{x} \neq \mathbf{x}'$; that is,

$$
\mathrm{cov}(Z_e(\mathbf{x}), Z_c(\mathbf{x}')) \mid Z_c(\mathbf{x}) = 0 \qquad \forall \mathbf{x} \neq \mathbf{x}'.
$$

To describe the dependences of the Gaussian processes Z_c and Z_d, we consider a squared exponential kernel of the form $k(\mathbf{x}, \mathbf{x}') = \sigma^2 r(\mathbf{x}, \mathbf{x}', \boldsymbol{\theta})$, parameterized with the vector $\boldsymbol{\theta}$ and the variance parameter σ^2. The correlations between two points, represented by the D-dimensional feature vectors $\mathbf{x}$ and $\mathbf{x}'$, are captured by $r(\mathbf{x}, \mathbf{x}'; \boldsymbol{\theta})$, as

$$
r(\mathbf{x}, \mathbf{x}', \boldsymbol{\theta}) = \exp \left\{ -\sum_{j=1}^{D} \theta_j (x_j - x_j')^2 \right\}. \tag{6.7}
$$

With the squared exponential correlation kernel, the resulting data-driven portion of the co-variance matrix for the two-level co-kriging model is

$$
\begin{aligned}
\mathbf{K} &= \mathrm{cov}(\mathbf{X}, \mathbf{X}) \\
&= \begin{pmatrix} \mathrm{cov}\{Z_c(\mathbf{X}_c), Z_c(\mathbf{X}_c)\} & \mathrm{cov}\{Z_c(\mathbf{X}_c), Z_e(\mathbf{X}_e)\} \\ \mathrm{cov}\{Z_e(\mathbf{X}_e), Z_c(\mathbf{X}_c x)\} & \mathrm{cov}\{Z_e(\mathbf{X}_e), Z_e(\mathbf{X}_e)\} \end{pmatrix} \\
&= \begin{pmatrix} \mathrm{cov}\{Z_c(\mathbf{X}_c), Z_c(\mathbf{X}_c)\} & \mathrm{cov}\{Z_c(\mathbf{X}_c), \rho Z_c(\mathbf{X}_e) + Z_d(\mathbf{X}_e)\} \\ \mathrm{cov}\{\rho Z_c(\mathbf{X}_e) + Z_d(\mathbf{X}_e), Z_c(\mathbf{X}_c)\} & \mathrm{cov}\{\rho Z_c(\mathbf{X}_e) + Z_d(\mathbf{X}_e), \rho Z_c(\mathbf{X}_e) + Z_d(\mathbf{X}_e)\} \end{pmatrix} \\
&= \begin{pmatrix} \sigma_c^2 \, r_c(\mathbf{X}_c, \mathbf{X}_c) & \rho \sigma_c^2 \, r_c(\mathbf{X}_c, \mathbf{X}_e) \\ \rho \sigma_c^2 \, r_c(\mathbf{X}_c, \mathbf{X}_e) & \rho^2 \sigma_c^2 \, r_c(\mathbf{X}_e, \mathbf{X}_e) + \sigma_d^2 \, r_d(\mathbf{X}_e, \mathbf{X}_e) \end{pmatrix}.
\end{aligned}
$$

We note that $r(\mathbf{x}, \mathbf{x}') = r(\mathbf{x}', \mathbf{x})$, and since there are two correlations r_c and r_d, we need to estimate two hyper-parameter vectors $\boldsymbol{\theta}_c$ and $\boldsymbol{\theta}_d$, plus σ_c^2 and σ_d^2, and the scaling parameter ρ. One way we can do this by using the statistical independence of the Gaussian processes Z_c and Z_d and finding maximum likelihood estimates (MLEs) for the associated means (μ_c and μ_d), variances (σ_c^2 and σ_d^2), hyper-parameters ($\boldsymbol{\theta}_c$ and $\boldsymbol{\theta}_d$) and the scaling parameter ρ. Analytical expressions exist for MLEs for the means and variances, obtained by maximizing the natural logarithm of the two concentrated likelihood functions numerically to solve for $\boldsymbol{\theta}_c$, $\boldsymbol{\theta}_d$, and ρ. For our two-level example, Forrester et al. gives these details. Succinctly, the likelihood function is the probability of the data given the model. In the present case, it is a multivariate Gaussian. The sets of hyperparameters are independent variables, and the cheap data is independent of the expensive data. Accordingly, we can find the hyperparameters of the cheap data by taking the natural logarithm of its Gaussian contribution to likelihood function and then solving the set of equations that result from setting the derivatives of this function with respect to the parameters equal to zero. After we find these values, we substitute them into the natural logarithm of the Gaussian for the difference. We now only need to consider the values of the targets at data points common to both the cheap and expensive data. Then, after we maximize the remaining Gaussian contribution to the likelihood, the location of the maximum yields the predictive mean and the covariance about this estimate yields the variance of the mean.

With the hyper-parameters estimated, the co-kriging predictions of mean ($\mu_e^\star$) and variance ($\sigma_e^{2\star}$) for the high-fidelity model, that is, the more expensive function, at a new point $\mathbf{x}^\star$ in the feature space, are

$$
\begin{aligned}
\mu_e^\star &= \hat{\mu} + \mathbf{k}^{\mathrm{T}} \mathbf{K}^{-1}(\mathbf{y} - \hat{\mu}\mathbf{1}) && (6.8) \\
\sigma_e^{2\star} &= \hat{\rho}^2 \hat{\sigma}_c^2 + \hat{\sigma}_d^2 - \mathbf{k}^{\mathrm{T}} \mathbf{K}^{-1} \mathbf{k}, && (6.9)
\end{aligned}
$$

where $\hat{\mu} = \mathbf{1}^{\mathrm{T}} \mathbf{K}^{-1} \mathbf{y} / \mathbf{1}^{\mathrm{T}} \mathbf{K}^{-1} \mathbf{1}$, $\mathbf{1}$ is a vector whose components are all one, and the caret symbol over ρ, σ_c, and σ_d denotes the MLEs of these parameters. The column vector $\mathbf{k}$ (that is, the block matrix boarding $\mathbf{K}$ to the right and from below) is given by

$$
\mathbf{k} = \begin{pmatrix} \hat{\rho} \hat{\sigma}_c^2 r_c(\mathbf{X}_c, \mathbf{x}^\star) \\ \hat{\rho}^2 \hat{\sigma}_c^2 r_c(\mathbf{X}_e, \mathbf{x}^\star) + \hat{\sigma}_d^2 r_d(\mathbf{X}_e, \mathbf{x}^\star) \end{pmatrix}.
$$

$r_c(\mathbf{X}_c, \mathbf{x}^\star)$, for instance, denotes a column vector of correlations between the cheap data $\mathbf{X}_c$ and the new point $\mathbf{x}^\star$ for which predictions are being made.

It is also important to note that the framework discussed here assumes that there is no noise or uncertainty in the high-fidelity data and Equation (6.8) is an interpolator of the high-fidelity model; that is, predictions at the known high-fidelity points will always coincide with the training values with vanishing variance. The technique extends to the case with noise at the high fidelity level. When noise is present, the predictions of the high-fidelity points no longer coincide with the training values. We also comment that sequential co-kriging can replace sequential kriging in the efficient global optimization method (EGO) in Chapter 4 and can be used in multi-objective optimization in Chapter 5 to create a multi-fidelity approach to these types of problems.

6.3 RECURSIVE CO-KRIGING

It is easier to motivate recurisive co-kriging then to explain it. For the two fidelity level co-kriging discussion just concluded, the predictive mean and variance both depend on the inverse of the matrix $\mathbf{K}$. The order of this matrix is the total number of cheap and expensive data points, $n_c + n_e$. Inverting this matrix becomes times consuming, if not prohibitive, as the total number of data becomes large. This number can become large quickly particularly as the number of fidelity levels becomes large. The recent work of Le Gratiet and Garnier [4, 5] mitigated this problem by showing that any co-kriging scheme with s-levels of variable-fidelity information sources decouples and equivalently reformulates in a recursive fashion as s kriging problems. The maximum order of the matrix needing inversion thus becomes that of the size of the largest data set at any level of fidelity. The difficulty in explaining their result is the mathematical expressions are a bit unwieldy. Instead, we first state the classical auto-regressive approach for s-levels and universal co-kriging and then state the change made by Le Gratiet and Garnier.

Under the assumption that we have results from multiple computer simulations or experiments, arranged in order of increasing fidelity and modeled by Gaussian processes $Z_t(\mathbf{x})$ for $t = 1, 2, \ldots, s$, the auto-regressive model for $t = 2, \ldots, s$ is

$$
\begin{aligned}
Z_t(\mathbf{x}) &= \rho_{t-1}(\mathbf{x}) Z_{t-1}(\mathbf{x}) + \delta_t(\mathbf{x}) \\
\rho_{t-1}(\mathbf{x}) &= \mathbf{g}_{t-1}^T(\mathbf{x}) \boldsymbol{\alpha}_{t-1},
\end{aligned}
$$

where $\delta_t(\mathbf{x})$ is modeled by a Gaussian process with mean $\mathbf{f}_t^T(\mathbf{x}) \boldsymbol{\beta}_t = \sum_i^{n_{f_t}} f_{t,i}(\mathbf{x}) \beta_{t,i}$ and variance $\sigma_t^2 r_t(\mathbf{x}, \mathbf{x}')$. It is assumed that $Z_1(\mathbf{x}) = \delta_1(\mathbf{x})$. Additionally, it is assumed that $Z_{t-1}(\mathbf{x})$ and $\delta_t(\mathbf{x})$ are statistically independent and $\operatorname{cov}(Z_t(\mathbf{x}), Z_{t-1}(\mathbf{x}') \mid Z_{t-1}(\mathbf{x})) = 0$ for all $\mathbf{x} \neq \mathbf{x}'$.

In this more general statement of the auto-regressive scheme, the previous scaling parameter ρ is replaced by a level-dependent function of $\mathbf{x}$, that is, it is replaced by $\mathbf{g}_t^T(\mathbf{x}) \boldsymbol{\alpha}_t = \sum_i^{n_{g_t}} g_{t,i}(\mathbf{x}) \alpha_{t,i}$. This function, as the position dependent mean, is described by user-defined functions $g_{t,i}(\mathbf{x})$ and $f_{t,i}(\mathbf{x})$ with regression and trend parameters $\boldsymbol{\alpha}_t$ and $\boldsymbol{\beta}_t$ that we need to fix at each level in addition to the hyper-parameters in the kernels.

Le Gratiet and Garnier changed the above to

$$Z_t(\mathbf{x}) = \rho_{t-1}(\mathbf{x})\tilde{Z}_{t-1}(\mathbf{x}) + \delta_t(\mathbf{x})$$
$$\rho_{t-1}(\mathbf{x}) = \mathbf{g}_{t-1}^T(\mathbf{x})\boldsymbol{\alpha}_{t-1},$$

where $\tilde{Z}_{t-1}(\mathbf{x})$ is a Gaussian process whose distribution is conditioned on the data and the values of the hyper, trend, and regression parameters up to and including level t. They then showed that the Gaussian process $Z_t(\mathbf{x})$ samples the normal distribution $\mathcal{N}(\mu_{Z_t}(\mathbf{x}), s_{Z_t}^2(\mathbf{x}))$ and that at level s, $\mu_{Z_s}(\mathbf{x})$ and $s_{Z_s}^2(\mathbf{x})$ equal the predictive mean and variance at level s of the classical auto-regressive scheme. The implication is that we can build up to the highest level of fidelity in steps where at each step our covariance matrix is sized by the amount of data available at the current stage of fidelity. An additional implication is that each step produces a kriging model for that level of fidelity, enabling comparisons of how and where adding data of higher fidelity changes the lower fidelity predictions. The build-up is not based on applying kriging to a given level independently of the other levels. The build-up is recursive in the sense that the predictive mean and covariance at level t depend on those quantities at level $t - 1$. A gain occurs because at level t we only need to invert the covariance matrix built for the data at that level.

6.4   MULTI-FIDELITY LEARNING FOR MATERIALS DESIGN AND DISCOVERY

As discussed in the previous chapters, a standard practice in materials informatics and machine learning is to utilize available training data that is generated at a consistent level of experiment or theory to build surrogate models of material properties. For instance, standard machine learning (ML) models utilizing Gaussian process regression, kernel ridge regression, random forests regression, or neural networks based models are heavily employed to estimate a diverse range of materials properties. A common theme underlying all such commonly employed machine learning models is that they are built using training data obtained from a single, consistent source. Thus, we can refer to such standard models to as single-fidelity models. However, upon exploring materials data from a practical standpoint, it is very common to encounter cases where a target property is estimated through multiple sources (experiments and physical models) varying both in accuracy and cost. Generally, the accuracy of the measurement or simulation is proportional to its cost. A few representative examples of such cases, belonging to different length scales, are outlined in Figure 6.1.

The first example highlights a hierarchy of increasingly complex approximations to the electronic exchange-correlation (xc) effects within the framework of DFT. Originally put forward by Perdew, the Jacob's ladder of density functional approximations to the xc energy [6] connects the "Hartree-Fock world" to the "heaven of chemical accuracy," with each rung representing an increased level of complexity as well as a higher accuracy in electronic xc effects and an increased computational cost. For instance, functionals represented by the two lower most

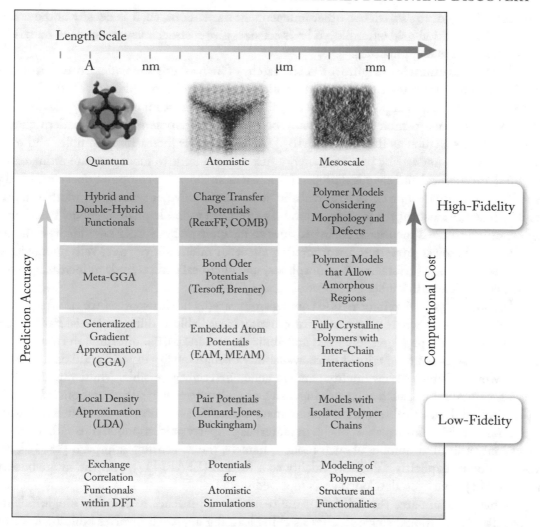

Figure 6.1: Selected prototypical examples in computational materials science where multi-fidelity information is typically available. Relevant length scales for each example are identified in the top panel and associated cost-accuracy tradeoff is highlighted across varying fidelities.

rungs account only for local density (local density approximation, LDA) and gradient (generalized gradient approximation, GGA) of the local density to the xc energy [7, 8]. Meta-GGA functionals include higher order gradient terms. Hybrid functionals (such as HSE06 and PBE0) rely on mixing exact electronic exchange from HF calculations to the approximate correlation effects within DFT [9]. More advanced approaches, such as random phase approximations (RPA) or double hybrid functionals, incorporate virtual/unoccupied orbitals [10]. Finally, in addition to the complexity of the xc effects, other independent dimensions, such as density of the reciprocal space integration grid or employed basis set size, give cost-accuracy tradeoffs in material's electronic structure computations.

The second example in Figure 6.1 is a hierarchy of increasingly sophisticated atomistic potentials available for finite-temperature atomistic simulations of molecules and solids. Moving from simple analytical pair potentials and (modified) embedded atom method ((M)EAM) [11, 12] based potentials to more complex many-body bond order potentials and modern charge transfer potentials (such as ReaxFF, COMB) [13, 14] captures increasingly complex and realistic details of the underlying potential energy surfaces and leads to more accurate estimates of structural, mechanical and chemical properties, albeit at an increased computational cost. Lastly, if one aims to estimate a property of a "realistic" amorphous polymer system, yet another multi-fidelity hierarchy exists at the mesoscale level. In this case, a simple physical model based on a single, periodic, one-dimensional chain provides an inexpensive first-order low-fidelity estimate of the target chemical properties. Incorporating effects of interchain van der Walls interactions, polymer morphology (crystalline vs. amorphous) and chemical defects leads to increasingly sophisticated models with higher predictive accuracy.

In addition to having information coming from varying fidelity sources, we also frequently encounters situations where only partial information is available at different levels. Particularly for materials science problems, the need and ability to include partial information from multiple fidelity sources is rather critical since availability of high quality experimental data is often limited, while volumes of low fidelity data are generally easily accessible. Multi-fidelity information fusion methods, such as co-kriging, can be highly effective in such situations. However, despite their high practical relevance, development of machine learning-based models that handle the multi-fidelity information fusion in materials property prediction remains rather scarce. In fact, only a limited number of such models, namely, the Δ-learning approach [15, 16], regression models explicitly using low-fidelity as a feature (LFAF) [17], and the multi-fidelity co-kriging [3] have been developed and employed to-date.

Table 6.1 compares these different multi-fidelity approaches to standard single-fidelity models. The conventional single-fidelity machine learning approach involves building a surrogate model that maps easily accessible material's attributes (features) to a target property computed or measured at a single level of fidelity. The Δ-learning and LFAF approaches are closely related, though not equivalent, and can be considered as naive versions of the multi-fidelity co-kriging. The former aims at learning differences between the estimates of a target property, computed at two levels of fidelities: $\mathbf{y_c}$ (low-fidelity) and $\mathbf{y_e}$ (high-fidelity). The final high-fidelity estimate is made by adding the learned difference, that is, Δ, to the known low-fidelity value, that is, $y_e = y_c + f^\Delta(\mathbf{x}^*)$. On the other hand, the LFAF approach explicitly augments the feature set with the low-fidelity value to directly learn the corresponding high-fidelity estimate, that is, $\mathbf{y_e} = f^{\mathrm{LFAF}}(\{\mathbf{X}, \mathbf{y_c}\})$. Therefore, unlike the Δ-learning approach, in LFAF $\mathbf{y_c}$ explicitly constitutes one of the dimension in the multi-dimensional feature space, wherein the machine

Table 6.1: A comparative description of various single and multi-fidelity machine learning approaches. $f(\mathbf{X}) \rightarrow \mathbf{y}$ represents a machine-learned mapping function from the feature space $\mathbf{X}$ to the property space $\mathbf{y}$. The feature vector $\mathbf{X}$ comprised of a set of relevant features pre-selected based on domain-knowledge, while $\mathbf{y}_e$ and $\mathbf{y}_c$ represent the high- and low-fidelity estimates, respectively, for the target property of interest. See text for details.

Machine Learning Model Classes	Nature of the Material-Property Mapping	Relative Remarks
Conventional single-fidelity (SF) models	$f(X) \rightarrow y_e$	Lower accuracy, cheaper predictions
Δ-learning models	$f^{\Delta}(X) \rightarrow \Delta y;$ $\Delta y = y_e - y_c$	Higher accuracy, expensive predictions
Models with low-fidelity as a feature (LFAF)	$f^{LFAF}(X, y_c) \rightarrow y_e$	Higher accuracy, expensive predictions
Multi-fidelity co-kriging	$f^{CK}(X): \left\{ \begin{array}{c} \rightarrow y_c \\ \updownarrow \\ \rightarrow y_e \end{array} \right\} \rightarrow y_e$	Higher accuracy, cheaper predictions

learning-based interpolation is performed. Both Δ-learning and LFAF require the availability of the low-fidelity property estimates for each and every material sample for which a high-fidelity prediction is sought and therefore can be computationally demanding when faced with a vast chemical space exploration. The relatively advanced multi-fidelity co-kriging approach alleviates this issue by modeling the available data as two independent Gaussian processes, one for the low-fidelity data and one for the difference between the two fidelities, while explicitly taking into account the pair-wise correlations between features, low-fidelity data, and high-fidelity data [2], as detailed in Section 6.2. As a result, predictions using the multi-fidelity co-kriging approach can be made with or without the low-fidelity data. While each of the three multi-fidelity methods, in principle, accounts for more than two fidelities by resorting to a recursive formulation, multi-fidelity co-kriging is the most flexible, as it provides a pathway to combine different numbers of low- and high-fidelity data points during model training. This is not possible with Δ-learning and LFAF.

A practical example demonstrating the advantage of a multi-fidelity learning approach over standard single-fidelity-based approaches was recently presented by Batra and co-workers [18]. This study focused on dopant formation energy of various substitutional dopants in hafnia (HfO_2), a well-known high dielectric permittivity material which is currently under active investigation for its promising ferroelectric, piezoelectric and pyroelectric properties, and utilized a database of DFT-computed dopant formation energy of 42 different dopants across the Periodic Table, each computed in six different phases of HfO_2. These six HfO_2 phases included monoclinic (M) $P2_1/c$, tetragonal (T) $P4_2/nmc$, cubic (C) $Fm\bar{3}m$, orthorhombic (OA) $Pbca$, polar orthorhombic (P-O1) $Pca2_1$, and another polar orthorhombic (P-O2) $Pmn2_1$ phases. In order to carryout a comparative analysis of single-fidelity and different multi-fidelity approaches (namely, the LFAF, Δ-learning and co-kriging methods), the dopant formation energy dataset was constructed at two levels of fidelity by modifying DFT computational parameters, such as k-point sampling, plane wave basis set's cut-off energy, and extent of allowed geometric relaxations. The high-fidelity values were generated from fully converged computations, while the low-fidelity data employed choices of the computational parameters that enabled these calculations at a minimal possible computational expense.

Figure 6.2a presents the test performance of co-kriging approach for the prediction of the dopant formation energies as a function of number of low-fidelity (n_c) and high-fidelity (n_e) data points used during the training process. The lower region in the figure is blank, since the high-fidelity dataset was always required to be a subset of the low-fidelity dataset. There are three important observations: (1) The prediction error on unseen data (consistently evaluated over the 48 randomly-selected left-out test cases) decreases with an increase in the number of both the low- and high-fidelity data points, with the lowest RMSE being around 0.45 eV when $n_e = n_c = 200$. Sample parity plots in Figure 6.2b quantitatively showcase the improvement in performance of co-kriging model with increasing values of n_e and n_c. (2) The learning rate is more sensitive to the number of high-fidelity data points included in the model as compared to the low-fidelity ones. This seems reasonable since the high fidelity data naturally carries more relevant information critical for a model's predictive performance improvement. (3) Finally, a RMSE of ~0.66 eV is achieved in the top-left region of the plot with the only 50 high-fidelity and 200 inexpensive low-fidelity data points. In comparison, (c) the RMSE for $n_e = 50$ is around 1.8 eV with SF(GPR), and 0.75 eV for Δ(GPR) and LFAF(GPR) models, as depicted in Figure 6.2. This comparison between the Δ-learning and the LFAF approaches, however, is not fair: The CK approach as the former cases utilize additional low-fidelity information for test set as well. Nonetheless, the results suggest that the CK approach is particularly beneficial when exceedingly large amount of low-fidelity data is available in comparison to the high-fidelity data. Such scenarios are encountered when the cost of a low-fidelity estimate is much less than that of a high-fidelity estimate. Overall, this study demonstrated that the utilization of additional relevant information contained in the low-fidelity estimates improves the prediction performance of all three multi-fidelity approaches and their predictions are superior to traditional single-fidelity machine learning methods, such as Gaussian process regression. Among the three multi-fidelity approaches, co-kriging is the most efficient and flexible. While the prediction accuracy of the multi-fidelity approaches are comparable (Figure 6.2c), the co-kriging approach had a much lower cost than the Δ-learning and LFAF models.

As another example of a practical application of the multi-fidelity learning approach, Pilania et al. [19] successfully demonstrated efficacy of the co-kriging method for learning bandgaps

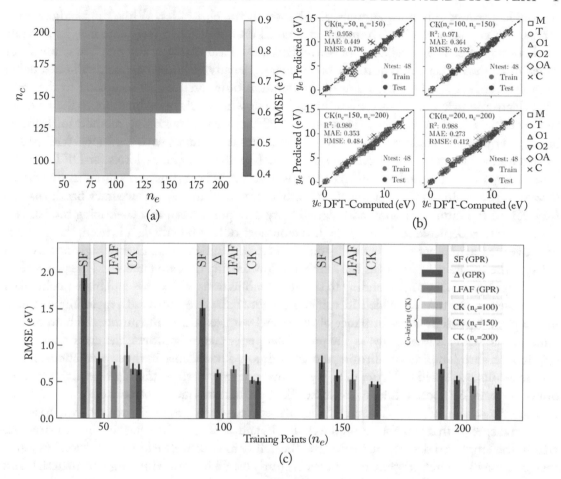

Figure 6.2: (a) Prediction accuracy of co-kriging models for the estimation of dopant formation energies in hafnia [18]. The co-kriging models were trained using different combinations of n_e high-fidelity and n_c low-fidelity data points. All results in (a) are for test sets consisting of 48 randomly selected points and averaged over 50 runs. Panels in (b) are example parity plots with varying training sets and a test set of 48 points. The different symbols indicate different HfO_2 phases. The performance is quantified by using different error measures: goodness of fit (R^2), mean absolute error (MAE), and root mean squared error (RMSE). (c) A comparison of the prediction accuracy of four different learning models as a function of number of high- and low-fidelity data points used in the training. Results for three flavors of Gaussian process regression (GPR) models are shown, namely, the standard single-fidelity (SF) model, the Δ-learning model, and the LFAF model. Cases where $n_e \not\subseteq n_c$ are omitted for the co-kriging approach. All results are averaged over 50 runs, with error bars representing one standard deviation. Adapted from Ref. [18].

of double-perovskite halide materials, also known as the elpasolites. While the bandgap often serves as a crucial screening parameter in rational design of functional materials, its accurate prediction is often a big challenge for conventional DFT calculations, and one frequently needs to resort to the higher rungs of the Jacob's ladder of density functional approximations. Higher rung calculations can be tremendously computation time intensive as compared to the standard electronic stricture calculations utilizing local or semi-local approximations to the xc functional. Pilania et al. successfully demonstrated efficacy of the co-kriging method for learning bandgaps of double-considered a set of 600 elpasolites compounds with semi-local (cheap) and hybrid exchange-correlation (expensive) functionals as the two fidelity levels for DFT calculations. They first computed all higher- and lower-accuracy bandgaps and subsequently studied how well the multi-fidelity optimization method predicted the higher-accuracy bandgaps as a function of the number of low- and high-fidelity data points used in a co-kriging model. Not surprisingly, as depicted in Figure 6.3a, if the number of low-fidelity data is fixed, the accuracy improves as the number of high-fidelity data is increased, and if the number of high-fidelity data is fixed, the accuracy improves as the number of low-fidelity data are increased. The parity plots shown in Figure 6.3b, corresponding to specific combinations of the low- and high-fidelity data points used in the trained models identified in Figure 6.3a, also showcase a rapid improvement in the test set prediction performance of the model with respect to the number of high fidelity data points. More interestingly, as the prediction performance improves, the error bars on the predictions shrink rapidly, confirming model's improved confidence in those predictions. Based on the results presented in Figure 6.3a, we see that one only needs ∼150 high-fidelity bandgaps out of the total 600 possibilities to make excellent predictions (mean absolute error of < 0.1 eV) on the entire chemical space. The target chemical space is described in Figure 6.3c, where various elemental species that can appear on the A, B, B', and X sites of the elpasolite crystal structure within the employed dataset are identified. The ability to effectively extract and readily use such insights, for any target property of interest, is clearly useful in both designing new materials and optimizing existing ones for a specific application.

Given that materials science is filled with examples of hierarchical data from high-throughput DFT computations and experiments with varying levels of accuracy, the above studies point to the promise of the multi-fidelity co-kriging approach for materials science problems. Since the multi-fidelity learning approaches are not restricted to any specific property, any particular method for data generation (for instance, to computational or experimental methods) or to any length scale (for example, electronic, atomistic, or mesoscale scales), it is easy to envision numerous examples where such a framework is applicable. The multi-fidelity information fusion implicitly extends the utility of low fidelity, high throughput calculations in a significant way and is trivially generalized to handle more than two levels of fidelities. Besides making higher fidelities estimates of one targeted property, the method is easily extended to predict multiple properties simultaneously. Used this way, it provides valuable surrogates for materials search problems seeking to identify possible candidates with multiple enhanced functionalities. This

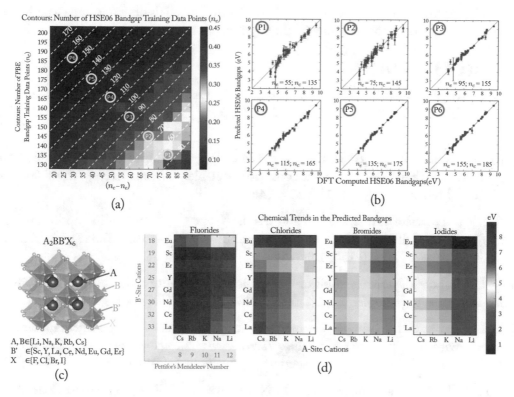

Figure 6.3: (a) Learning performance, quantified as mean absolute error in eV, of the multi-fidelity co-kriging model in predicting bandgaps of elpasolite compounds as a function of number of low- and high-fidelity data points, that is, n_c and n_e, used in the training. (b) Panels P1–P6 show representative parity plots comparing the high-fidelity bandgap predictions of the trained model and the DFT-computed high-fidelity bandgaps on the cross-validation set, a set of 50 left-out unseen data points, for selected pairs of n_c and n_e. The panel labels are also appropriately indexed in (a). (c) Crystal structure of $A_2BB'X_6$-type elpasolite compounds. The octahedral environment of B- and B'-site cations is shown explicitly and the chemical space relevant to this study is also depicted. (d) Analysis of chemical trends for the predicted bandgaps. The four panels (from left to right) classify different X-site chemistries for the elpasolite compounds. A- and B'-site elemental species are ordered according to the Pettifor's Mendeleev number and each panel depicts an average predicted bandgap for a given A- and B'-site chemistry and averaged over all B-site chemistries in the dataset. Adapted from [19].

capability should be valuable in multi-objective optimization problems [20, 21], discussed in the previous chapter. The Gaussian process nature of the method also makes it convenient for use in conjunction with other methods based on adaptive materials design, paving a pathway for advanced information-theoretic multi-fidelity optimization in materials science.

6.5   REFERENCES

[1] D. G. Krige. A statistical approach to some basic mine valuation problems on the Witwatersrand. *Journal of the Chemical, Metal, and Mining Society of South Africa*, 52:119, 1959. 135

[2] M. C. Kennedy and A. O'Hagan. Predicting the output from a complex computer code when fast approximations are available. *Biometrika*, 87:1, 2000. DOI: 10.1093/biomet/87.1.1 135, 137, 139, 145

[3] A. I. J. Forrester, A. Sóbester, and A. J. Keane. Multi-fidelity optimization via surrogate modelling. *Proc. of the Royal Society A*, 463(2088):3251, 2007. DOI: 10.1098/rspa.2007.1900 135, 138, 144

[4] L. L. Gratiet. Bayesian analysis of hierarchical multifidelity codes. *SIAM/ASA Journal of the Uncertainty Quantification*, 1(1):244, 2013. DOI: 10.1137/120884122 137, 141

[5] L. L. Gratiet and J. Garnier. Recursive co-kriging model for design of computer experiments with multiple levels of fidelity. *International Journal of Uncertainty Quantification*, 4(5), 2014. DOI: 10.1615/int.j.uncertaintyquantification.2014006914 137, 141

[6] J. P. Perdew, A. Ruzsinszky, J. Tao, V. N. Staroverov, G. E. Scuseria, and G. I. Csonka. Prescription for the design and selection of density functional approximations: More constraint satisfaction with fewer fits. *Journal of Chemical Physics*, 123(6):062201, 2005. DOI: 10.1063/1.1904565 142

[7] J. P. Perdew, K. Burke, and M. Ernzerhof. Generalized gradient approximation made simple. *Physical Review Letters*, 77(18):3865, 1996. DOI: 10.1103/physrevlett.77.3865 143

[8] R. O. Jones. Density functional theory: Its origins, rise to prominence, and future. *Reviews of Modern Physics*, 87(3):897, 2015. DOI: 10.1103/revmodphys.87.897 143

[9] J. Heyd, G. E. Scuseria, and M. Ernzerhof. Hybrid functionals based on a screened Coulomb potential. *Journal of Chemical Physics*, 118(18):8207, 2003. DOI: 10.1063/1.1564060 143

[10] X. Ren, P. Rinke, C. Joas, and M. Scheffler. Random-phase approximation and its applications in computational chemistry and materials science. *Journal of Materials Science*, 47(21):7447, 2012. DOI: 10.1007/s10853-012-6570-4 143

[11] S. M. Foiles, M. I. Baskes, and M. S. Daw. Embedded-atom-method functions for the fcc metals Cu, Ag, Au, Ni, Pd, Pt, and their alloys. *Physical Review B*, 33(12):7983, 1986. DOI: 10.1103/PhysRevB.33.7983 144

[12] M. I. Baskes. Modified embedded-atom potentials for cubic materials and impurities. *Physical Review B*, 46(5):2727, 1992. DOI: 10.1103/physrevb.46.2727 144

[13] T. P. Senftle, S. Hong, M. M. Islam, S. B. Kylasa, Y. Zheng, Y. K. Shin, C. Junkermeier, R. Engel-Herbert, M. J. Janik, and H. M. Aktulga. The ReaxFF reactive force-field: Development, applications and future directions. *NPJ Computational Materials*, 2:15011, 2016. DOI: 10.1038/npjcompumats.2015.11 144

[14] T. Liang, T.-R. Shan, Y.-T. Cheng, B. D. Devine, M. Noordhoek, Y. Li, Z. Lu, S. R. Phillpot, and S. B. Sinnott. Classical atomistic simulations of surfaces and heterogeneous interfaces with the charge-optimized many body (COMB) potentials. *Materials Science and Engineering R*, 74(9):255, 2013. DOI: 10.1016/j.mser.2013.07.001 144

[15] B. Huang, N. O. Symonds, and O. A. von Lilienfeld. Quantum machine learning in chemistry and materials. *Handbook of Materials Modeling: Methods: Theory and Modeling*, p. 1, Springer, Berlin, 2018. DOI: 10.1007/978-3-319-42913-7_67-1 144

[16] R. Ramakrishnan and O. A. von Lilienfeld. Many molecular properties from one kernel in chemical space. *CHIMIA International Journal of Chemistry*, 69(4):182, 2015. DOI: 10.2533/chimia.2015.182 144

[17] J. Lee, A. Seko, K. Shitara, K. Nakayama, and I. Tanaka. Prediction model of band gap for inorganic compounds by combination of density functional theory calculations and machine learning techniques. *Physical Review B*, 93(11):115104, 2016. DOI: 10.1103/physrevb.93.115104 144

[18] R. Batra, G. Pilania, B. P. Uberuaga, and R. Ramprasad. Multifidelity information fusion with machine learning: A case study of dopant formation energies in hafnia. *ACS Applied Materials and Interfaces*, 2019. DOI: 10.1021/acsami.9b02174 145, 147

[19] G. Pilania, J. E. Gubernatis, and T. Lookman. Multi-fidelity machine learning models for accurate bandgap predictions of solids. *Computational Materials Science*, 129:156, 2017. DOI: 10.1016/j.commatsci.2016.12.004 146, 149

[20] A. Mannodi-Kanakkithodi, G. Pilania, T. D. Huan, T. Lookman, and R. Ramprasad. Machine learning strategy for accelerated design of polymer dielectrics. *Scientific Reports*, 6:20952, 2016. DOI: 10.1038/srep20952 150

[21] A. Mannodi-Kanakkithodi, G. Pilania, R. Ramprasad, T. Lookman, and J. E. Gubernatis. Multi-objective optimization techniques to design the Pareto front of

organic dielectric polymers. *Computational Materials Science*, 125:92, 2016. DOI: 10.1016/j.commatsci.2016.08.018 150

CHAPTER 7

Some Closing Thoughts

In this final chapter, we present three, somewhat independent, short essays on topics that provide an important perspective on the nature of machine learning, propose a potential future machine learning opportunity for materials design and discovery, and make some comments about the use of machine learning in experimental material science. In the first instance, we discuss machine learning form the perspective of probability theory and Bayes's Theorem. In the second, continuing the theme of methods prefixed with "multi-", we note the existence of multirelational methods and how they could change our perspective of how data is organized. Finally, as theorists, who are grateful for the experimental support some of our work has received, we give a few examples of how machine learning methods is impacting the way experiments are being executed.

7.1 THE BAYESIAN PERSPECTIVE

In several places, we stated that a particular technique is Bayesian. By this we mean the method was developed and is stated in terms of probability theory and in particular uses Bayes's Theorem,

$$P(A|B) = \frac{P(B|A)P(A)}{P(B)}, \tag{7.1}$$

where A and B are two sets of events, $P(A)$ and $P(B)$ are their individual probabilities, plus $P(A|B)$ and $P(B|A)$ are their conditional probabilities. For data analysis [1], we usually write this theorem as

$$P(\text{Model}|\text{Data}) = \frac{P(\text{Data}|\text{Model})P(\text{Model})}{P(\text{Data})}. \tag{7.2}$$

$P(\text{Model}|\text{Data})$ is called the posterior distribution (the probability of the model after taking the data into account); $P(\text{Data}|\text{Model})$ is the likelihood function (the probability of the data given the model); and $P(\text{Model})$ is called the prior (what we know about the model before the data is taken into account). Knowing the posterior constitutes knowing the complete probabilistic solution to the problem. $P(\text{Data})$ is called the evidence. It is the normalization of the posterior; that is,

$$P(\text{Data}) = \int_{\text{Model}} P(\text{Model}|\text{Data}) = \int_{\text{Model}} P(\text{Data}|\text{Model})P(\text{Model}). \tag{7.3}$$

In data-fitting analysis, the logarithm of Bayes's Theorem is generrally used

$$\log[P(\text{Model}|\text{Data})] \propto \log[P(\text{Data}|\text{Model})] + \log[P(\text{Model})] \tag{7.4}$$

with the logarithm of the evidence (that is, the logarithm of the normalization of the posterior) is dropped. Generally unstated is that much of machine learning is equivalent to maximizing the log posterior with respect to the parameters of the model with the evidence ignored. From this point of view, we see that machine learning, instead of seeking the complete solution, generally settles for estimates of the location of the maximum of the posterior (the mean) and variance about the mean (the width of the posterior distribution about its maximum). The log likelihood is the cost, loss, or utility function for the problem. An important note is that the familiar least-squares fitting of a model to data is equivalent to maximizing

$$\log[P(\text{Model}|\text{Data})] \propto \log[P(\text{Data}|\text{Model})]. \tag{7.5}$$

Adding the log of the prior "regularizes," that is, makes robust, the fitting of the model to the data.

In machine learning, the connection to Bayes's Theorem is generally not noted as little insight is gained by doing so. Typical prior information about the model includes assumptions about such things as smoothness, that is, mathematical information. What one can see is the potential of the Bayesian perspective for adding more specific physics-based assumptions for the prior, such as bounds on model parameters, correlations in the data to be favored by the model, etc. Doing so would give the machine learning a more specific material science character.

The evidence is usually ignored because it is hard to evaluate. Gaussian processes are an exception. Typically, steepest decent approximations, that is, Gaussian approximations to the integrands of various integrals, is often made that allows the evidence's approximation analytically. Monte Carlo methods are increasingly being used to estimate the evidence numerically [1]. What is too involved to discuss is that the inclusion of the evidence gives an Occam's razor penalty for a model with more parameters that the data warrants [1]. The Bayesian point of view, that is, including a prior probability, makes valid a formal approach to model selection that becomes an alternative to the heuristic of cross-validation. From (7.4) [1]:

$$\frac{P(\text{Model}_1|\text{Data})}{P(\text{Model}_2|\text{Data})} = \frac{P(\text{Data}|\text{Model}_1)}{P(\text{Data}|\text{Model}_2)} \times \frac{P(\text{Model}_1)}{P(\text{Model}_2)} \tag{7.6}$$

from which the more probable model for the given data is inferable.

7.2   TENSORS NOT TABLES OF DATA

While the Bayesian approach is a natural and often touted way to embed prior domain knowledge, we now note another approach that focuses on what we can say *a priori* about the data instead of what we can say about the model. This approach is called multi-relational learning. We are unaware of any application of this approach in materials design and discovery.

The canonical example of multi-relational learning is: Suppose we have lists of the names of past U.S. presidents and vice-presidents (the entities) and partial knowledge of the their political party, who they were president of, and who they were vice-president for (three relations).

Mindful that in U.S. history, we have had more political parties than Democrat or Republican and the vice-presidents were not always of the same party as the president, the task is to build a model that predicts the parties of each president and vice-president.

How might one approach this problem with a machine learning method? All methods we discussed start with a data matrix that typically has materials as its rows and features as its columns. From this matrix, we extract subsets of materials and features and then transform the reduced matrix for use in a specific machine learning method. The matrices we input to these methods might be the "distance matrix" between materials, covariances among features, etc. For use in a multi-relational learning method, we could instead express the data differently [2–5] whereby the data is mapped to, or simply assembled as, multiple matrices of the same dimensions where each matrix groups data according to different relations based on our prior domain knowledge. The data matrix $\mathcal{D}_{mn}$ maps to a tensor χ_{ijk}. In multi-relational learning the tasks are to choose the entities and state the relations.

In materials science, one or both entities defining the dimensions of matrices χ_k for relation k most naturally would be a material, but the two entities could also be a pair of features, two quantities which are not features, etc. Presently, data tables mix different types of features together. Instead, we collect separate tables in which in each the data is naturally associated, say those describing crystal structure, alloy concentration, functionality, etc. The relations might be functional, F_k(value of ith entity) = value of jth entity, or logical, (ith entity, kth predicate, jth entity) with $\chi_{ijk} = 1$ if a relation exists or 0 if it is does not. As implied by our "presidential" example, multi-relational learning does not require values for all entity pairs in the tables. Our machine learning can now return information not only about relations within a table and but also about those among the tables. Additionally, its models "fill-in" the missing entries.

7.3   MACHINE LEARNING IN EXPERIMENT

Just as advances in numerical methods and computer power are allowing the high-throughput calculations of properties of a combinatorial large number of possible materials, advances in combinatorial synthesis and characterization are allowing the exploration of large parameter spaces for rapid screening of possible new materials. For example, sputtering thin film can synthesize composition gradients in three or more elements of promising compositions for targeted properties, such as lattice parameters and local co-ordination [6]. The in-depth bulk synthesis and analysis on just a few select compounds are performed. This approach is a down-selection strategy rather than an active learning one discussed in Chapter 4. In future, we see active learning becoming possible: Advances in measurement techniques will allow more high throughput experiments [7], and then the concomitant ability to make on-the-fly decisions as to what next to synthesize and test by using, for example, the methods of Chapter 4 to save cost and time. Feature sets will include processing conditions, such as laser power, travel speeds, and cooling rates, and with such sets it will become possible to predict and control the resultant material microstructure.

Big data, up to 100 TBs per sample, are presently being generated at such user facilities, such as the Advanced Photon Source at Argonne and the Linac Coherent Light Source at the Stanford Linear Accelerator Center. A challenging task is reconstructing the real-space macrostructural image from the data. For example, in High Energy Diffraction Microscopy, a forward model needs to be run at every finite element to find the orientation of the "grain" that matches the Bragg spots on the detector pattern [8]. Besides requiring substantial computational resources, this task causes a time delay in performing the next experiment. However, machine learning and optimization methods, and in particular convolution neural nets (CNNs) [9], provide a vehicle to construct models, depending on the numbers of neurons and hidden layers, that directly fit large amounts of data to hundreds and thousands of parameters. In this CNN activity, the speed-up relies on constructing machine-learned surrogate models whose predictions eliminate the need for complex calculations associated with the original problem. Much current debate is about the merits of deep learning models, especially because they often appear as "black-boxes." Using information theory and the renormalization group, Lin et al. [10] discussed why deep learning works so well and how "cheap learning" may be crafted with far fewer parameters.

7.4   REFERENCES

[1] D. S. Sivia and J. Skilling. *Data Analysis: A Bayesian Tutorial.* Oxford University Press, Oxford, 2006. 153, 154

[2] L. Getoor and B. Taskar Eds. *Introduction to Statistcal Relational Learning.* MIT Press, Cambridge, MA, 2007. DOI: 10.7551/mitpress/7432.001.0001 155

[3] L. De Raedt. *Logical and Relational Learning.* Springer, New York, 2008. DOI: 10.1007/978-3-540-68856-3

[4] T. G. Kolda and B. W. Bader. Tensor decompositions and applications. *SIAM Review,* 51:455, 2009. DOI: 10.1137/07070111x

[5] M. Nickel, V. Tresp, and H.-P. Kriegel. A three way model for collective learning on multi-relational data. In L. Getoor and T. Scheffer, Eds., *Proc. of the 28th International Conference on Machine Learning,* p. 809, ACM, Bellvue, WA, 2011. 155

[6] H. Koinuma and I. Takeuchi. Combinatorial solid-state chemistry of inorganic materials. *Nature Materials,* 3:429, 2004. DOI: 10.1038/nmat1157 155

[7] A. G. Kusne, T. Gao, A. Mehta, L. Ke, M. C. Nguyen, K.-M. Ho, V. Antropov, C.-Z. Wang, M. J. Kramer, C. Long, and I. Takeuchi. On-the-fly machine-learning for high-throughput experiments: Search for rare-earth-free permanent magnets. *Scientific Reports,* 4:6367, 09 2014. DOI: 10.1038/srep06367 155

[8] S. F. Li and R. M. Suter. Adaptive reconstruction method for three-dimensional orientation imaging. *Journal of Applied Crystallography*, 46(2):512, 2013. DOI: 10.1107/s0021889813005268 156

[9] N. Lubbers, T. Lookman, and K. Barros. Inferring low-dimensional microstructure representations using convolutional neural networks. *Physical Review E*, 96:052111, 2017. DOI: 10.1103/physreve.96.052111 156

[10] H. W. Lin, M. Tegmark, and D. Rolnick. Why does deep and cheap learning work so well? *Journal of Statistical Physics*, 168(6):1223, 2017. DOI: 10.1007/s10955-017-1836-5 156

APPENDIX A

Basic Notions of Probability Theory

While the concept of probability is something about which we all have a good intuitive feeling, in this Appendix, we will express this intuition more formally. At its conclusion, we will have the background to appreciate better machine learning methods from the point of view of learning probabilistically.

If asked to define *probability*, we all would most likely say it is the expected number of times a particular event occurs divided by the number of events that occurred. This notion is basically a operational definition of how to measure the probability of an stochastic event, usually based on the fiction of performing many identical experiments and simply recording the number of times the stochastic event of interest is observed. If the experiment is flipping a coin, for example, we expect that if the coin is fair and the number of flips is large, half the time we would observe "heads."

Here, we will also consider probability as a measure of our degree of belief that an event will occur. This notion fits our natural tendency to reason probabilistically: If we believe that event A is more likely than B and B is more likely than C, then it seems that A is more likely than C is valid conclusion more so than simply a belief. Probability theory allows us to reason in such a seemingly consistent fashion but not without controversy.

The controversy is between the "frequentist" and the "Bayesian" viewpoints of probability and statistics. A text such as the one by Sivia and Skilling [1] gives a short discussion of both viewpoints. The point of contention is the degree to which "belief" makes the analysis subjective. Our intent here is to give a short background to the Bayesian (belief) approach to probabilistic learning as this approach allows us in many cases to state and model our assumptions and then to reason consistently. Our conclusions ultimately are testable by the extent to which they are consistent with the data and with experiment.

The probability $P(A)$ of an event A is a non-negative number lying in the closed interval $[0, 1]$. If the number is 1, then the event is certain; if 0, then it is impossible. One can reason consistently and logically from two relations. The first is if the N events in the set $\{A_1, A_2, \ldots, A_N\}$ exhaust all possibilities and are mutually exclusive, then

$$\sum_A P(A) \equiv \sum_{i=1}^{N} P(A_i) = 1. \tag{A.1}$$

The second is if we have two events A and B, then the probability of their jointly occurring $P(A, B)$ satisfies

$$P(A, B) = P(A|B)P(B),\tag{A.2}$$

where $P(A|B)$ is the conditional probability that A occurs provided B occurs. In many respects, this last equation defines what we mean by a *conditional probability*. It also has other implications. If the occurrence of A is independent of B, then we should have $P(A|B) = P(A)$. From the above, we can then infer that $P(A, B) = P(A)P(B)$. In this situation, we say that A and B are statistically independent. If instead we had defined *statistical independence* of A and B as $P(A, B) = P(A)P(B)$, then the last equation says that $P(A|B)$ must equal $P(A)$. Logical consistency is thus illustrated.

We must also have the $P(A, B) = P(B, A)$. The last equation implies

$$P(A|B)P(B) = P(B|A)P(A),$$

which is often written as

$$P(A|B) = \frac{P(B|A)P(A)}{P(B)}.\tag{A.3}$$

This relation is called *Bayes's Theorem*. It allows us to "flip" conditional probabilities.

An often understated but important point is all probabilities are in fact conditional. Assigning a probability in a coin flip that $P(A)$ is the probability of "heads" is usually done so on the condition that the coin and flip are "fair." As the conditional probability for A given B is thus a probability, from A.1) we can then write $\sum_A P(A|B) = 1$, and then from (A.2), we find that

$$\sum_A P(A, B) = P(B),$$

This summation is called the *marginalization* of A. It removes A from the statement of the probability. From (A.3), we can also write that

$$P(B) = \sum_A P(B|A)P(A).$$

This relation allows us to interpret $P(B)$ as the normalization constant of $P(A|B)$.

We have been "overloading" the symbols A, B, etc., referring them as the events, random variables, or values assumed by the random variables. These are distinct concepts. In general, the context makes to which concept we are referring clear. We have also implied that the values of the random variables are discrete whereas in many applications they are continuous. Alternately, when continuous, we can then write (A.1) as

$$\int P(A)dA = 1$$

and replace the various summations by integrations.

When the variables are continuous, the concept of a *probability distribution function* (pdf) $p(A)$ or distribution for short is important. In terms of a distribution, the probability that the value of A is between a and b is

$$P(a < A < b) = \int_a^b p(A) dA.$$

A probability density satisfies $p(A) \geq 0$ and $\int p(A) dA = 1$.

In probabilistic reasoning and learning, the concepts of joint probabilities, marginalization, and "flipping" our statements of probability via Bayes's Theorem play a central role. Their use allows us to write such chains of equations as

$$P(A, C) = \sum_B P(A|B, C) P(B|C) = \sum_{B,D} P(A|B, C, D) P(D|B, C) P(B|C),$$

thereby enabling us the reduce a complicated chain of probabilities to a simpler one or to expand a simpler one into a more complicated chain for which we have several pieces we know how to model.

When our degrees of freedom are continuous, Gaussian distributions are particularly convenient. Marginalizing variables in a Gaussian produces another Gaussian. The product of two Gaussian distributions is also a Gaussian. These facts are what helps make Gaussian processes convenient. Before we discuss these processes in Appendix C, we first discuss in Appendix B some the basic algebra and calculus for Gaussian distributions.

APPENDIX B

Multivariate Gaussian Distributions

While various properties of univariate Gaussian functions are easy to derive, those of multivariate Gaussians require much more algebra. In machine learning, most often our Gaussians are multivariate. Below we state without derivation several useful properties of these distributions.

By a *multivariate Gaussian* we mean

$$\mathcal{N}(x|\mu, \Sigma) \equiv p(x|\mu, \Sigma) = \frac{1}{\sqrt{(\det(2\pi\Sigma)}} \exp\left[(x - \mu)^T \Sigma^{-1}(x - \mu)\right], \qquad \text{(B.1)}$$

where $x = (x_1, x_2, \ldots, x_D)^T$ is a D-dimensional vector of the independent variables. μ is a vector of the mean values of the x variables, and Σ is their $D \times D$-dimensional covariance matrix; that is,

$$\mu = \langle x \rangle, \quad \Sigma = \langle (x - \mu)(x - \mu)^T \rangle.$$

The angular brackets denote an expectation value with respect to the multivariate Gaussian (B.1).

Useful properties of multivariate Gaussians include [2, 3]

1. The product of two Gaussians is a Gaussian:

$$\mathcal{N}(x|\mu_1, \Sigma_1)\mathcal{N}(x|\mu_2, \Sigma_2) = \mathcal{N}(x|\mu, \Sigma)\frac{\exp(-\frac{1}{2}(\mu_1 - \mu_2)^T A^{-1}(\mu_1 - \mu_2)}{\sqrt{(\det(2\pi A)}}, \qquad \text{(B.2)}$$

   where the $N \times N$ matrix $A = \Sigma_1 + \Sigma_2$, the N vector $\mu = \Sigma_1 A^{-1}\mu_2 + \Sigma_2 A^{-1}\mu_1$, and the $N \times N$ matrix $\Sigma = \Sigma_1 A \Sigma_2$,

2. The marginalization of some degrees of freedom yields a Gaussian: If

$$p(\mathbf{z}) = \mathcal{N}(\mathbf{z}|\mu, \Sigma)$$

   with

$$\mathbf{z} = \begin{pmatrix} x \\ y \end{pmatrix}, \quad \mu = \begin{pmatrix} \mu_x \\ \mu_y \end{pmatrix}, \quad \Sigma = \begin{pmatrix} \Sigma_{xx} & \Sigma_{xy} \\ \Sigma_{yx} & \Sigma_{yy} \end{pmatrix},$$

   where the dimensions of x and y may differ, then

$$p(x) = \mathcal{N}(x|\mu_x, \Sigma_{xx}), \qquad \text{(B.3)}$$

3. The conditional probability density of the partition of the variables is a Gaussian:

$$p(\boldsymbol{x}|\boldsymbol{y}) = \mathcal{N}(\boldsymbol{x}|\boldsymbol{\mu}_x + \Sigma_{xy}\Sigma_{yy}^{-1}(\boldsymbol{y} - \boldsymbol{\mu}_y), \Sigma_{xx} - \Sigma_{xy}\Sigma_{yy}^{-1}\Sigma_{yx}. \tag{B.4}$$

We note that $\Sigma_{yx} = \Sigma_{xy}^T$.

4. The transformation of Gaussian variables by a Gaussian process yields another Gaussian process. If

$$\boldsymbol{y} = A\boldsymbol{x} + \boldsymbol{\eta},$$

where $\boldsymbol{x}$ and $\boldsymbol{\eta}$ are statistically independent and both sampled from Gaussians, that is,

$$\boldsymbol{x} \sim \mathcal{N}(\boldsymbol{\mu}, \Sigma), \quad \boldsymbol{\eta} \sim \mathcal{N}(\boldsymbol{\mu}_\eta, \Sigma_\eta).$$

Then the marginal $p(\boldsymbol{y}) = \int p(\boldsymbol{x}, \boldsymbol{y})d\boldsymbol{x} = \int p(\boldsymbol{y}|\boldsymbol{x})p(\boldsymbol{x})d\boldsymbol{x}$ is another Gaussian,

$$p(\boldsymbol{y}) = \mathcal{N}(\boldsymbol{y}|A\boldsymbol{\mu} + \boldsymbol{\mu}_\eta, A\Sigma A^T + \Sigma_\eta). \tag{B.5}$$

APPENDIX C

Gaussian Processes

A *Gaussian process* is a collection of random variables such that any finite collection of them has a multivariate Gaussian distribution. In terms of machine learning algorithms, Gaussian processes are Bayesian methods. These methods do not target a best fit but instead compute a posterior (probability) distribution over functions (models). The latter provides not only a complete solution to the problem but also an avenue to the quantification of the uncertainty in the predictions of the model. The significance of assuming Gaussian distributions is that we can exactly perform many of the required algebraic operations and integrations by executing simple linear algebra operations on vectors and matrices (Appendix B).

Suppose we have a training set of n observations $\mathcal{D} = \{(x_i, y_i), | i = 1, \ldots, n\}$ where an x_i is a vector of D features. Alternately, we can write this as $\mathcal{D} = (X, y)$ where X is a $D \times n$ matrix whose columns are the feature vectors and y is a vector collecting the labels (that is, targets) y_i. What we seek is the conditional probability of the targets given the inputs. Even more specifically, if x_* is a new input, we seek $p(y_*|x_*)$, the probability distribution of the new target y_*.

We are doing *supervised learning*, which is the problem of learning models from the empirical data. The two main cases of supervised learning are *regression* and *classification*. In regression, the y_i are continuous variables; for classification, they are discrete. We first discuss Gaussian processes for regression, and then we discuss classification. Our treatment abbreviates those of Rasmussen and Williams [2] and Barber [3].

C.1 REGRESSION

We start by following the somewhat traditional route to motivating Gaussian processes by first considering the construction of a Bayesian linear regression model. In the simplest case, we have

$$f(x) = x^T w, \quad y = f(x) + \varepsilon, \tag{C.1}$$

where x is the input vector, w is a vector of weights (the parameters of the model), f is a function value at x, and y is the observed value of the target. We assume the difference between the observed and function values are independently and identically distributed Gaussian noise ε with zero mean and variance σ_n^2; that is,

$$p(\varepsilon) = \mathcal{N}(\varepsilon | 0, \sigma_n^2).$$

Our initial objective to construct the posterior distribution. The noise and the linear model give rise to the likelihood function

$$
\begin{aligned}
p(y|X, w) &= \prod_{i=1}^{N} p(y_i|\mathbf{x}_i, w), \\
&= \prod_{i=1}^{N} \frac{1}{\sqrt{(2\pi)}\sigma_n} \exp\left(-\frac{|y - \mathbf{x}_i^T|^2}{2\sigma_n^2}\right), \\
&= \mathcal{N}(y|X^T w, \sigma_n^2 I),
\end{aligned}
$$

where $|\mathbf{x} - y|$ is the Euclidean distance between $\mathbf{x}$ and y. We now assume a zero mean Gaussian prior with a covariance matrix Σ_w for the weights,

$$
p(w) = \mathcal{N}(w|0, \Sigma_w).
$$

Now Bayes's Theorem tells us that

$$
p(w|y, X) = \frac{p(w, y|X,)p(w|X)}{p(y|X)}.
$$

The values of the weights however are not conditional on the input data matrix X so we replace $p(w|X)$ by the Gaussian $p(w)$. Using the properties of multivariate Gaussians, we can show that

$$
p(w|X, y) = \mathcal{N}(w|\sigma_n^{-2} A^{-1} X y, A^{-1}), \quad A = \sigma_n^2 X X^T + \Sigma_w^{-1}.
$$

To make a prediction for a test case $\mathbf{x}_*$, we average over all possible parameter values, weighted by their posterior probability. If $f_* = f(\mathbf{x}_*)$, then

$$
\begin{aligned}
p(f_*|\mathbf{x}_*, X, y) &= \int p(f_*|\mathbf{x}_*, w)p(w|X, y)d w \quad &\text{(C.2)} \\
&= \mathcal{N}(f_*|\sigma_n^{-2}\mathbf{x}_*^T A^{-1} X y, \mathbf{x}_*^T A^{-1}\mathbf{x}_*). \quad &\text{(C.3)}
\end{aligned}
$$

The mean of a Gaussian is also the most probable value of the random variable. Using this as the solution, we have

$$
\begin{aligned}
\hat{f}_* &= \sigma_n^{-2}\mathbf{x}_*^T A^{-1} X y, \quad &\text{(C.4)} \\
\text{cov } \hat{f}_* &= \mathbf{x}_*^T A^{-1}\mathbf{x} \quad &\text{(C.5)}
\end{aligned}
$$

with $\hat{f}_*$ as our predictive value. We note that the predictive variance matrix is a quadratic form of the test input with the posterior covariance matrix.

Next, we increase the complexity of the linear model by introducing a function $\phi(\mathbf{x})$ that maps the D-dimensional input vector $\mathbf{x}$ into an N-dimensional space of functions. For example, if $D = 1$, $\phi(x)$ might be $(1, x, x^2, \ldots, X^N)$. Our model is now

$$
f(\mathbf{x}) = \phi(\mathbf{x})^T w. \quad \text{(C.6)}
$$

If $\Phi(X)$ is a matrix whose columns are the $\boldsymbol{\phi}(\boldsymbol{x})$, then the analysis of this model follows that of the simpler linear model (C.1), and we can use the results of that analysis by simply replacing $\boldsymbol{x}$ with $\boldsymbol{\phi}(\boldsymbol{x})$. Thus, from (C.2)

$$p(f_*|\mathbf{x}_*, X, \boldsymbol{y}) = \mathcal{N}(f_*|\boldsymbol{\phi}(\mathbf{x}_*)^T A^{-1} \Phi(X)\boldsymbol{y}, \boldsymbol{\phi}(\mathbf{x}_*)^T A^{-1} \boldsymbol{\phi}(\boldsymbol{x}_*)). \tag{C.7}$$

With considerable algebra, we can rewrite the above as

$$p(f_*|\mathbf{x}_*, X, \boldsymbol{y}) = \mathcal{N}(f_*|\boldsymbol{\phi}(\boldsymbol{x}_*)^T (K + \sigma_n^2 I)^{-1} \boldsymbol{y},$$
$$\boldsymbol{\phi}(\boldsymbol{x}_*)^T \Sigma_w \boldsymbol{\phi}(\boldsymbol{x}) - \boldsymbol{\phi}(\boldsymbol{x}_*)\Sigma_w \Phi(X)(K + \sigma_n^2 I)^{-1} \Phi(X)^T \Sigma_w \boldsymbol{\phi}(\boldsymbol{x}),$$

where the matrix $K = \Phi^T \Sigma_w \Phi$. This rewriting reveals the higher dimensional space always enters in the form $\Phi \Sigma_w \Phi$, $\boldsymbol{\phi}_*^T \Sigma_w \Phi$ or $\boldsymbol{\phi}_*^T \Sigma_w \boldsymbol{\phi}_*$. Each of these expressions in turn are of the quadratic form $\boldsymbol{\phi}(\boldsymbol{x})^T \Sigma_w \boldsymbol{\phi}(\boldsymbol{x}')$. Because Σ_w is a positive-definite matrix, it has a Cholesky decomposition into the product of its square roots. Accordingly, we can define the *covariance function (kernel)* by

$$k(\boldsymbol{x}, \boldsymbol{x}') = \boldsymbol{\phi}(\boldsymbol{x})^T \Sigma_w \boldsymbol{\phi}(\boldsymbol{x}') = \boldsymbol{\psi}(\boldsymbol{x})^T \boldsymbol{\psi}(\boldsymbol{x}'),$$

where $\boldsymbol{\psi} = \Sigma^{\frac{1}{2}} \boldsymbol{\phi}$; that is, we can transform the covariance defining the prior for the weights out of the problem with a redefinition of the basis functions.

In machine leaning, when an algorithm is defined solely in terms of the dot product of two vectors, such as the above, the "*kernel trick*" is often invoked. This trick is simply replacing the dot-product with a continuous function $k(\boldsymbol{x}, \boldsymbol{x}')$ of the variables $\boldsymbol{x}$ and $\boldsymbol{x}'$. Whereas we were lead to the kernel by using a finite-dimensional basis in our modeling, we could start with the kernel and work back to the basis. For many commonly used kernels, the corresponding basis is infinite dimensional, hence the advantage of the kernel trick. This function space represents a space of covariance values, not the covariance matrix we would normally calculate from the input data.

We now come to *Gaussian processes*. Our function $f(\boldsymbol{x})$ is regarded as something sampled from a Gaussian defined by a mean $\mathbf{m}(\boldsymbol{x})$ and covariance function $k(\boldsymbol{x}, \boldsymbol{x}')$. The expectation of the mean is typically chosen to be zero, leaving $\text{cov}(f(\boldsymbol{x}), f(\boldsymbol{x}')) = k(\boldsymbol{x}, \boldsymbol{x}')$. A Gaussian process is thus defined by the covariance function. In these processes, we still can use the various relations for multivariate Gaussian that enable use to exactly express our predictions. We now have

$$\begin{bmatrix} \boldsymbol{y} \\ f_* \end{bmatrix} = \mathcal{N}\left(\begin{bmatrix} \mathbf{0} \\ \mathbf{0} \end{bmatrix} \middle| \begin{bmatrix} \mathbf{0} \\ \mathbf{0} \end{bmatrix}, \begin{bmatrix} K(X, X) + \sigma_n^2 I & K(X, X_*) \\ K(X_*, X) & K(X_*, X_*) \end{bmatrix} \right), \tag{C.8}$$

where $K(X, X_*)$ is the $n \times 1$ matrix of covariance pairs between the training data and test point. $K(X, X)$, $K(X_*, X)$ and $K(X_*, X_*)$ are similarly defined. Our predictive distribution becomes

$$p(f_*|\boldsymbol{x}_*, X, \boldsymbol{y}) = \mathcal{N}(\hat{f}_*, \text{cov}(\hat{f}_*)), \tag{C.9}$$

where

$$\hat{f}_* \equiv K(X_*, X)[K(X, X) - \sigma_n^2 I]^{-1} y, \tag{C.10}$$
$$\text{cov}(\hat{f}_*) = K(X_*, X_*) - K(X_*, X)[K(X, X) - \sigma_n^2 I]^{-1} K(X, X_*). \tag{C.11}$$

There are a large number of choices for the covariance function. Most have been well studied with respect to their smoothness and the length scales over which they correlate the data. Several choices include the *generalized squared exponential*

$$k(x, x') = \sigma_0 \exp\left(-\frac{1}{2}\sum_{i=1}^{D}\lambda_i(x_i - x_i')^2\right)$$

and the *exponential-sine square*

$$k(x, x') = \exp\left(-2[\sin(\tfrac{\pi}{p}d(x, x'))]/\ell]^2\right),$$

where $d(x, x')$ is the Euclidian distance between x and x'. Other functions generally have multiple parameters, such as the $(\sigma_0, \lambda_1, \ldots, \lambda_D)$ and (p, ℓ) in the above, that need learning. These parameters are called *hyperparameters*. The Bayesian approach to setting the hyperparameters is to maximize the log of the marginal likelihood (evidence)

$$\log p(y|X) = -\tfrac{1}{2}y^T(K + \sigma_n^2 I)y - \tfrac{1}{2}\det(K + \sigma_n^2 I) - \tfrac{n}{2}\log 2\pi. \tag{C.12}$$

The kernel K contains the dependencies on the hyperparameters.

C.2   CLASSIFICATION

Classification is a supervised learning task in which the targets assume only discrete as opposed to continuous values. The application of Gaussian processes to this task differs from the application to regression by having some distributions other than Gaussians. In addition, the concept of *latent variables* is used. In short, we predict the classes from these variables. The values of these variables we learn from the data. Schematically,

$$p(c|x) = \int p(c, y|x)dy = \int p(c|x, y)p(y|x)dy \rightarrow \int p(c|y)p(y|x)dy.$$

In the last step, we assume the class mapping from the data x and the latent variable y does not depend on the data. A Gaussian process for $p(y|x)$ will induce the mapping of the data to the latent variable. In the end, the latent variables are integrated out of the problem. We will discuss the basics for using Gaussian processes for classification. Details needed for implementation are left to standard texts [2, 3].

Suppose we have a training set of n observations $\mathcal{D} = \{(x_i, c_i), i = 1, \ldots, n\}$ where an x_i is the input vector of D features and the c_i are the values of the classes associated with this data.

Alternatively, we can write this as $\mathcal{D} = (X, c)$ where X is a $D \times n$ matrix whose columns are the vector inputs and c is a vector collecting the outputs (targets) c_i. The most common case is that of binary classification. Here, the c_i may have the values of True or False, Yes or No, etc. that map onto two integers, for example, 1 and 0. The analysis we present applies to the binary case as well as cases where there are more than two classes.

We seek the conditional probability of the targets given the inputs. The target with the highest probability is generally taken as the predicted class. Even more specifically, if x_* is a new input, we seek $p(c_*|x_*)$, the probability distribution of the new target c_*. We start by writing

$$
\begin{aligned}
p(c_*|x_*, X, c) &= \int p(c_*, y_*|x_*, X, c)dy_* \\
&\propto \int p(c_*|y_*)p(y_*|x_*, X, c)dy_*.
\end{aligned}
$$

In the first equation, we introduce the latent variable y_*, which we do not know and which does not have any specific connection to the data. In the second, we are using Bayes's Theorem and leaving out the conditional distribution that is independent of the integration variables. Now,

$$
\begin{aligned}
p(y_*|x_*, X, c) &\propto p(y_*, c|x_*, X), \\
&= \int p(y_*, y, c|x_*, X)dy, \\
&= \int p(c|y)p(y_*, y|x_*, X)dy,
\end{aligned}
$$

where we invoked the relation between joint and conditional probabilities, introduced the latent y variables, and then exploited the independence of the class mappings from the input data. We now need to relate the latent variables to the input.

We start by observing that

$$
p(y_*, y|x_*, X, c) \propto p(y_*, y, c|x_*, X) \propto p(y_*|y, x_*, X)p(y|c, X). \tag{C.13}
$$

In the above the term $p(y_*|y, x_*, X)$ lacks class labels and is simply a conditional Gaussian. We now approximate $p(y|X, c)$ by a Gaussian $q(y|X, c)$. Thus we have

$$
p(y_*, y|x_*, X, c) \approx p(y_*|y, x_*, X)q(y|c, X).
$$

We then marginalize the y from this Gaussian to find a Gaussian distribution for the y_* and hence ultimately find the predictive distribution $p(c_*|y_*)$. Explicit expressions for the predictors depend on whether the classification involves two or more classes [2, 3].

For binary classification problems, the distribution mapping the latent variable to class labels is usually taken to be a *sigmoid function*, that is, an "s-shaped" function. For example, for $c = \{0, 1\}$,

$$
p(c|y) = \sigma[(2c - 1)y], \quad \sigma(x) = (1 + e^{-x})^{-1}.
$$

$\sigma(-x) = 1 - \sigma(x)$, ensuring $\sum_c p(c|y) = 1$.

C.3   REFERENCES

[1] D. S. Sivia and J. Skilling. *Data Analysis: A Bayesian Tutorial.* Oxford University Press, Oxford, 2006. 159

[2] C. E. Rasmussen and K. J. Williams. *Gaussian Processes for Machine Learning.* MIT Press, Cambridge, 2006. DOI: 10.7551/mitpress/3206.001.0001 163, 165, 168, 169

[3] D. Barber. *Bayesian Reasoning and Machine Learning.* Cambridge University, Cambridge, 2010. 163, 165, 168, 169

Authors' Biographies

GHANSHYAM PILANIA

Ghanshyam Pilania is a scientist in the Materials Science and Technology Division at Los Alamos National Laboratory (LANL). He received a B.Tech. in Metallurgical and Materials Engineering from Indian Institute of Technology Roorkee, India in 2007, followed by a Ph.D. in Materials Science and Engineering from University of Connecticut, Storrs, in 2012. His four year postdoctoral work was supported by a LANL Directors' postdoctoral fellowship award and an Alexander von Humboldt postdoctoral fellowship at the Fritz Haber Institute of the Max Planck Society. His current research interests broadly include developing and applying high throughput electronic structure and atomistic methods to understand and design functional materials, with a particular focus on targeted materials design and discovery using materials informatics and machine learning based techniques.

PRASANNA V. BALACHANDRAN

Prasanna V. Balachandran is currently an Assistant Professor with a joint appointment in the Department of Materials Science and Engineering and Department of Mechanical and Aerospace Engineering in University of Virginia (UVA). He earned his Bachelors' Degree in Metallurgical Engineering from Anna University, India in 2007 and a Ph.D. in Materials Science and Engineering from Iowa State University, in 2011. Prior to joining UVA in December 2017, he spent three and half years as a postdoctoral research associate in the Theoretical Division at Los Alamos National Laboratory (LANL), and two years as a postdoctoral research associate at Drexel University. His research interests are interdisciplinary spanning diverse areas such as crystal symmetry, first-principles-based density functional theory calculations, and information science methods for accelerating the design and discovery of new materials.

JAMES E. GUBERNATIS

James E. Gubernatis did his thesis work at CaseWestern Reserve University under the direction of P.L. Taylor and his post-doctoral work at Cornell University under the mentorship of J.A. Krumhansl. After his post-doctoral work, he joined the technical staff of the Los Alamos National Laboratory where he stayed until his recent retirement. During his career, his main research interests were Quantitative Nondestructive Evaluation using the scattering of elastic waves, Quantum Monte Carlo methods and their applications to interacting electron systems, and the application of machine learning methods to the design and discovery of materials. He is a past chair of the Division of Computational Physics of the American Physical Society and the Commission on Computational Physics of the International Union of Pure and Applied Physics. He is also a Fellow of the American Physical Society.

TURAB LOOKMAN

Turab Lookman obtained his Ph.D. from Kings College, London and was Professor of Applied Mathematics at the University of Western Ontario, Canada until 1999. He was at Los Alamos National Laboratory for almost 20 years until 2019. His interests and expertise lie in aspects of Computational Materials Science, Condensed Matter Physics, including soft matter, and aspects of nonlinear dynamics and chaos theory. His focus since 2012 has been on the application of machine learning methods to materials design and discovery. Many of the methods discussed in this book were utilized in 2013–2016 in a highly successful project that led to the synthesis of new alloys and ceramics. He was honored as a Laboratory Fellow in 2017 and is a Fellow of the American Physical Society.

Printed in the United States
by Baker & Taylor Publisher Services